Studies in Consciousness / Russell Targ Editions

S ome of the twentieth century's best texts on the scientific study of consciousness are out of print, hard to find, and unknown to most readers; yet they are still of great importance, their insights into human consciousness and its dynamics still valuable and vital. Hampton Roads Publishing Company—in partnership with physicist and consciousness research pioneer Russell Targ—is proud to bring some of these texts back into print, introducing classics in the fields of science and consciousness studies to a new generation of readers. Upcoming titles in the *Studies in Consciousness* series will cover such perennially exciting topics as telepathy, astral projection, the after-death survival of consciousness, psychic abilities, long-distance hypnosis, and more.

STUDIES IN CONSCIOUSNESS

Russell Targ Editions

Experiments in Mental Suggestion

L. L. Vasiliev

HAMPTON ROADS
PUBLISHING COMPANY, INC.

Hampton Roads Publishing Company, Inc.
1125 Stoney Ridge Road
Charlottesville, VA 22902

434-296-2772
fax: 434-296-5096
e-mail: hrpc@hrpub.com
www.hrpub.com

If you are unable to order this book from your local
bookseller, you may order directly from the publisher.
Call 1-800-766-8009, toll-free.

Library of Congress Catalog Card Number: 2002100963
ISBN 1-57174-274-3
10 9 8 7 6 5 4 3 2 1
Printed on acid-free paper in Canada

Contents

Foreword

Readers of this book will be amazed by the sophisticated and dramatic psychic experiments conducted to test extrasensory perception by L. L. Vasiliev, who was a pioneering Russian physiologist and psychologist. He transmitted telepathic instructions to create hypnotic sleep in willing (and sometimes unwilling) subjects, conducted ESP studies over hundreds of miles, and sealed participants in large lead and iron containers to shield against electromagnetic leakage. His accomplishments in experimental design included use of probability statistics, methods for evaluating successful ESP, and the use of shields, mechanical randomizing, and images as targets. His ingenuity is obvious, and his honest reporting of negative findings as well as stunning successes is an indication of his integrity. All this was in the 1920s and 1930s in Russia, where there was a climate of hostility to anything smacking of the non-physical and the farther ranges of the human mind. This book is a valuable record of work that is little known and is seldom given the recognition it deserves. Contemporary research on psi (the professional term for the varieties and processes of extrasensory perception) can still learn from Vasiliev's work and ideas.

This book by Leonid Leonidovich Vasiliev (1891–1966) first appeared in its original Russian edition (1962) under the title *Eksperimental'nye Issledovaniia Myslennogo Vnusheniia*. An English translation was published in 1963 as *Experiments in Mental Suggestion*. This present publication is from a slightly revised edition which appeared in 1976.

Vasiliev's research in parapsychology was marked by creativity, determination, and careful conceptualization. It is also striking for its courage, since his conclusion that the electromagnetic theory of telepathy was not confirmed by his research conflicted with the Soviet mandated worldview of materialism.

The central theme that runs through the book is a testing of the theory that the transmission of telepathic suggestions (and other forms of extrasensory perception) are via electromagnetic (EM) waves. Vasiliev places this in context for each study with a summary of parapsychology theory and research from Russia and other countries. His own original position was that the process of mental suggestion had a physicalistic basis, and he believed that once the phenomena could be reproduced in the laboratory, then their physical nature could be explored.

One of the unusual features of his work (at least for Western researchers) is that most of his telepathic experiments were conducted with the percipients in a hypnotic trance. Hypnotic induction and suggestions at a distance have been claimed since the early days of Mesmerism. In the 1799 "Dissertation by F. A. Mesmer, Doctor of Medicine, on His Discoveries," Mesmer (1980) asserted that the magnetic force that he postulated could work at a distance. The practitioners of Mesmerism and its later development as hypnosis reported effects such as mentally inducing somnambulism close up and far away *(il sommeil à distance)* and "community of sensation," in which the hypnotized percipient reacted to deliberately produced physical sensations of the hypnotist, such as touch, pain, taste, and smells, in some cases when the hypnotist was in another room. In Vasiliev's early studies, suggestions were given telepathically (i.e., mentally) to the subject to carry out a behavior (e.g., raise your right leg at the knee), to experience intense emotions, to go into hypnotic sleep and to wake up, and to reproduce drawings. These were successful in varying degrees. Vasiliev found that for him the most reliable of these effects were the suggestions to sleep and to wake, and most of the EM research was carried out using sleeping and waking induced by mental suggestion as the indication of telepathy. This does not refer to ordinary sleep, but hypnotic sleep, a trance state. A description by Julius Ochorovicz, early Polish psychologist and psychic researcher, is included in the book's appendix. Within two to five minutes, he writes, "her vision becomes blurred, her eyelids begin to flutter rapidly until finally the eyeball is concealed underneath the lid; her chest begins to heave, she behaves as if she were falling ill; she sighs and falls backwards into a deep sleep."

These phenomena were reported throughout Great Britain (e.g., by Drs. Esdaile and Elliotson) and Europe, including Russia and Poland. Some percipients were subjected to careful (for those days) experimental tests which appeared to establish paranormal receptivity, but they also revealed that some percipients were experiencing hyperacute sensory awareness or were perpetrating fraud. In France in the 1850s, physicians Pierre Janet and Joseph Gibert conducted experiments with a talented hypnotic subject Léonie B., a patient of the latter. Gibert found that he could mentally command Miss B. to go into a trance, and even carry out specific actions. Janet learned to do the same. Some of the distances between the hypnotist and subject were from one house to another, a distance of more than fifteen hundred feet (three hundred meters). The landlady kept records of when Miss B. dropped into sleep and sometimes the researchers would secretly observe her from outside. One series of the sleep suggestions, which were given at random times, succeeded in nineteen out of twenty-five trials. In another series, fifteen out of twenty-one trials were successful. The English psychical researcher F. W. H. Myers took part in some of the experiments and discussed them in his book *Human Personality and Its Survival of Bodily Death* (Myers 1904).

In Russia, an 1818 book on Mesmerism by D. Velianski reported somnambulism through mental commands. In the later 1800s there was a revival of interest in hypnotism, which was claimed to produce healing effects and telepathic and clairvoyant awareness. Meanwhile in Poland, Ochorovicz carried out trials of mental suggestion of behavior (Rise, go to the piano, take a box of matches, bring it to me . . .). He traveled to France, witnessed the work of Janet and Gibert, and was impressed enough by one experiment to be persuaded there was a causal connection between the act of will and an effect produced at a distance. Many reports of distant hypnosis are in Eric J. Dingwall's authoritative four volume work on nineteenth-century hypnosis, *Abnormal Hypnotic Phenomena* (Dingwall 1967–1968). Vasiliev was familiar with these accounts of mental hypnotic influence. He comments on some of them in the text and reproduces European and Russian reports in the appendices. However, in the last half of the nineteenth century, the apparent paranormal effects of hypnosis were being claimed and appropriated by spiritualism, and this association had a dampening effect on the interest in hypnosis. It was by treating the telepathic suggestions as psychological and physiological phenomena, subject to accepted methods of research, that Vasiliev and others were able to bring the research back into credibility, even though he was using hypnosis, itself controversial, as the facilitator of psi.

While telepathic suggestion was soon ignored in the West after the turn of the 1900s, it continued in the Soviet countries. As late as the 1960s, biophysicist Yuri Kamensky conducted a telepathic experiment from Moscow to Leningrad (five hundred miles) to Karl Nikolaev, who was connected to physiological monitors with the attempt to measure a biological response to the telepathic transmission. Feeling that negative emotions were transmitted more reliably, Kamensky imagined that he was strangling his partner. Nikolaev felt suffocated and his EEG showed drastic changes. In another test Kamensky imagined he was physically beating the percipient, who fell out of his chair in pain. The doctors who were monitoring his reactions were concerned for his life, according to Larissa Vilenskaya (1984). Nikolaev was not formally hypnotized, but he said that for psychic functioning he took about half an hour to put himself into a state of relaxation (Ostrander and Schroeder 1970). At the time of the East-West Cold War, Vilenskaya says also that the Soviets were attempting telepathic suggestions directed at the political leaders of the West, hoping to confuse their minds and political actions (Vilenskaya 1984). Vasiliev expresses regrets in his final chapter that the telepathic hypnotic method for sleeping and waking is no longer used in other countries' research, and comments that its development was a great achievement of the Soviet researchers.

The Russian work appears to assume that the sender should have the ability to concentrate attention and send it with a force of will, and the percipient must be talented and in a receptive state. British, European and U.S. research has focused more on the conditions of the receiver of the psi information, and more on information than behavior. I am inclined to interpret these different approaches to the hypnotic relationship as partially due to different cultural patterns toward authority, power, ego control, and compliance. This has also been the trend in hypnosis theory and practice. The will of the hypnotist is rarely emphasized, and his or her role is seen now as a facilitator of a hypnotic state. (This has even been carried by some to a more extreme position that even an altered state is not necessary to hypnotic responsiveness.) It might be productive now to revisit the idea of hypnotic transmission and the hypnotist's intentionality, and possible behavioral responses to distant suggestions.

A provocative discussion of the psychological dynamics of this issue comes from the late Jule Eisenbud, U. S. psychiatrist and Freudian analyst. He reported that he was curious about the neglect (or avoidance, as he said) of the early work on distant suggestion by Janet, Esdaile, Elliotson and others, given the potential significance of the work. He suspected that such distant influence might be a reality, and the professional neglect of it might have some psychoanalytic undercurrents. Never one to avoid intriguing and slightly risky investigations, Eisenbud decided to try telepathic suggestion himself. Beginning with a good hypnotic subject, a truck driver, he willed him to call him long distance, however forgetting that he (Eisenbud) was in a place with no telephone, so of course nothing happened. A week later he mentally projected a suggestion during a hypnotic session for the driver to telephone him later at five P.M. that afternoon. Nothing happened that afternoon, but a few days later, the driver telephoned out of the blue to tell Eisenbud that he was going on vacation, distinctly an unusual call. Then Eisenbud received three postcards from the distant driver, plus another telephone call. Eisenbud reflected that perhaps the first suggestion of a long distance call was being carried out while the driver was at a long distance on vacation, and was a fusion of the two instructions to call. At a later time the analyst sent another mental message to the driver to phone, and within a few hours, the call came.

Eisenbud tried this with two more persons, a patient who was a deep hypnotic subject, and a friend from whom he had not heard in fifteen months. Both of them phoned within hours. In his account Eisenbud relates that over a year later he realized that he had totally dropped the subject, and his notes were skimpy, rather than his usual detailed reports. He concluded he was "in flight" from the implications of his casual experiments. He felt the successes had triggered infantile desires for power, for omnipotence, for control over others, and had thus aroused powerful intrapsychic defenses against these desires, barriers necessary for human society and species survival, and important for a balanced personality. He commented that such capabilities also raise the possibility of a genuinely potent death wish, and a resulting fear of wielding such power. The hypnotic subject, on his or her part, may also resist the will of the hypnotist. Vasiliev mentions this occurring with his participants, and Miss. B. sometimes washed her hands vigorously to keep from being put into sleep by Gibert's distant suggestions. True to form, no one in the Western parapsychology establishment to my knowledge has conducted further investigations of telepathic commands.

However, the same underlying process has morphed into two topics in contemporary parapsychology research. The first is Direct Mental Interaction with Living Systems (DMILS) a phrase coined by William Braud. The second is distant healing, also variously named spiritual, mental, and psychic healing.

The DMILS work might be considered a form of psychokinesis (PK), but it is specifically directed toward living systems, and it does not make any theoretical assumptions, as does the concept of PK. In DMILS the attempt is made to influence behavior of biological systems (gerbils on exercise wheels, fish swimming in a tank), physiology (e.g. human electrical skin response) or their consciousness (e.g., ability to concentrate, awareness of being stared at) (Braud 1993). Meticulous and carefully designed experiments have confirmed such distant effects, and meta-analyses support the statistical significance of the body of research (Schlitz and Braud 1997). The conclusion is that mental intention directed toward a participant can effect physical and mental changes not involving the known physical mechanisms, i.e., via psi. The human participants for these studies are not given hypnotic inductions. Some of the conditions involve a quieting or relaxed state of mind and body, and others do not. The goals of the DMILS influences are intentionally different from enforcing the will of the agent, and move in the direction of outcomes which are neutral or might be considered positive for the receiver, including, I presume, the non-human participants.

The second area, distant healing, is aligned even more directly with life enhancing values. While non-physical modes of healing are present in religions and in early Mesmerism and hypnosis, only recently have they become the subjects of serious research studies. Initially these were conducted principally by parapsychologists (for which they deserve personal and professional credit) and ignored or rejected in the arenas of medicine (Hastings, Fadiman, and Gordon 1980; LeShan 1974). As of this writing, this isolation is shifting, and studies of distant physical and psychological healing are being conducted in medical, physiological, psychological, and parapsychological disciplines, with careful designs involving double blind conditions, randomized trials, sophisticated statistical analyses, and other standard research methods (Byrd 1988; Sicher, Targ, Moore, and Smith 1998). There is no assumption that this mode of healing is a cure-all; rather there is the hope of finding where it can be effective, and further, what the conditions are that correlate with its results, such as kinds of illness, the techniques of the healers, time lines of change, and possible mechanisms. Analyses of the many studies that have been done up to now show statistically significant results: healing effects are produced by what this

book would term distant influence. These results do not occur in all participants or in all studies, but the overall significance is clear, and indicates that
they are real (Abbott 2000; Astin, Harkness, and Ernst 2000; Benor 1992;
Winstead-Fry and Kijek 1999). The techniques used include a variety of techniques and metaphysical models: prayers to God, therapeutic touch, spiritual
practice, subtle energy, meditation, laying on of hands, eclectic healers, and
many other modalities. There seem to be no indications in the data that one
particular technique of healing is more efficacious than others.

Thus, while the telepathic projection of the will has languished (or been
excluded) from current research, the process of distant influence has been
revived in more beneficent forms which further the metavalues of healing, harmony of systems, and cooperation. Such values are more congruent with the
experienced values of self actualization, transpersonal states, and mystical experiences, now subjects themselves of research and systematic exploration.

To return to one point regarding the hypnotic model used by Vasiliev, we
do not know from the book what form of hypnotic induction was used to train
the subjects. Given that the experiments began in the 1920s and in Russia, I
would deduce that the technique was a traditional authoritarian approach, in
which the hypnotist creates eye fixation, usually by staring into the subject's
eyes, and gives forceful commands to "sleep!" I recall observing a
Czechoslovakian psychologist in the 1960s hypnotizing a person and emphat-
ically saying, "My will overcomes yours." I was taken aback by this forceful
manner, but it appeared to be effective. With Vasiliev, there may also have been
the use of passes, slowly moving the hands down over the subject's body, from
head to feet. This technique is left over from the Mesmeric days, and may be
used with or without suggestions of sleep. While contemporary thinking
rejects the idea that the passes transfer "magnetic fluid," the procedure was
effective in inducing a trance. (It could be that there is actually electromagnetic
or other energy radiated by this process, which is like non-contact laying on of
hands, but I know of no research on this.) Passes were said to produce deep
states of hypnosis, called somnambulism (supposedly similar to sleep walking,
but actually not the same) or a plenary trance. These profound and deeper levels can be induced with verbal suggestions also, and can produce non-ordinary
experiences such as shifts in perception and identity, physiological changes,
and even unitive states such as those described in spiritual accounts (Tart
1969). These states may be psychologically and physically healing and restora-
tive to the person.

Vasiliev also used hypnosis in his work on transmission of mental images.
This is a familiar psi target in current research, where images and designs have
been used in forced choice (e.g., card decks) and free response (e.g., drawings)
studies. The state of hypnosis has been found to consistently improve ESP scores,
regardless of what suggestions are given or even if suggestions are not given. This
has led into the exploration of body and mind conditions that enhance ESP

responses, characterized as psi-conducive states. These include physiological relaxation, mental quieting, inward attention, and a non-concentrative alertness (Braud and Braud 1974; Honorton 1977)—hypnosis, meditation, relaxation, and sensory reduction. Studies using such states show enhanced psi functioning.

Vasiliev found that the telepathic suggestion to sleep or to wake was the most reliable technique to use. Its effect was easily observed and it had a high rate of success. He could then begin to systematically test whether the signal was carried by electromagnetic waves. His successive experiments are detailed in the book. He shielded the transmitter and/or the percipient in Faraday chambers (electrically shielded rooms). The telepathic instructions still worked. He put them in separate rooms. The effects continued. Eventually he sealed them in iron and lead covered boxes. The image of a hypnotized subject, sitting in a totally dark and electrically shielded lead box with a round metal door at the top floating in a seal of mercury, and rhythmically squeezing a rubber ball to indicate wakefulness, is a unforgettable one, not easily matched in parapsychology. The telepathic suggestions still were effective.

Then Vasiliev placed the transmitter and receiver at longer distances, including a fifteen hundred mile separation between Sebastapol and Leningrad, and still the percipient went into hypnotic sleep on command. While they are less electronically sophisticated than would be designed at present, the experiments nevertheless provide sufficient evidence that the blocking of EM radiation does not significantly affect the telepathic transmission, a conclusion that is still current. However, he does not conclude that EM waves are excluded as the physical basis of telepathy, he simply says that the shielding of the EM waves did not affect the telepathic reception, a description appropriate as an honest scientist and circumspect for his political milieu.

At this writing the thought is that electromagnetic waves of any frequency are not likely to be the carriers of psi signals, though they may affect the physical reception of the message by the brain itself. This may also be the case with other forms of energy such as the Earth's magnetic field and stellar radiation, which have been correlated with changes in psi functioning. Almost all contemporary research finds, along with Vasiliev, that electrical shielding and distance have no statistically significant effect on experimental results, though the Russian notes that biological mechanisms may alter the effects of distance by amplifying low signals, or the message may be transmitted even at attenuated levels through information bits. On the other hand, the EM theory finds difficulty in explaining precognition, in which there is psi awareness of events that occur later in time, i.e., in the future.

Attention is now turning to quantum processes as possible physical bases for psi. Perhaps I should mention that Mesmer asserted that the nervous system was in continuity with a subtle fluid which could independently extend without air or ether to unlimited distances and could immediately contact the internal senses of another individual.

Vasiliev's experiments are landmarks. They were thought out carefully and systematically, and conducted with attention to detail. The results were solid, as indicated by more recent research which essentially has produced the same findings.

The final chapter is a thoughtful reflection on the dynamics of telepathy, theoretical issues, and comments on the research of others. It is worth study. His impressions about the nature and features of telepathy are insightful, and can still be pursued today.

Arthur Hastings
Institute of Transpersonal Psychology
Palo Alto, California, USA

Note

The introduction by Anita Gregory includes text and a table giving possible corrections to the 1976 English edition (of which this is a reprint). As the pages have been reset for this printing, the changes will be found as follows:

Table Number	Page This Printing
5	72–73
6	75–76
7	77
8	78
9	80
10	81
13	90
Page 106	65

References

Abbott, Neil C. (2000). Healing as a therapy for human disease: A systematic review. *Journal of Alternative and Complementary Medicine,* 6(2). 159–169.

Astin, J., Harkness, E., & Ernst, E. (2000). The efficacy of "distant healing": A systematic review of randomized trials. *Annals of Internal Medicine.* 132, 903–910.

Benor, D. (1992). *Healing Research—Research in Healing.* Vol. I. Munich, Germany: Helix Verlag.

Braud, W. (1993). On the use of living target systems in distant mental influence research. In L. Coly and J. D. S. McMahon (Eds.). *Psi Research Methodology: A Re-Examination* (pp. 149–188). New York: Parapsychology Foundation.

Braud, L. W., and W. G. Braud. (1974). Further studies of relaxation as a psi-conducive state. *Journal of the American Society for Psychic Research,* 68, 229–245.

Byrd, R. (1988). Positive therapeutic effects of intercessory prayer in a coronary care unit population. *Southern Medical Journal,* 81(7), 826–829.

Dingwall, E. (Ed.) (1967–1968). *Abnormal Hypnotic Phenomena: A Survey of Nineteenth-Century Cases.* Vol. I-IV. New York: Barnes and Noble.

Eisenbud, J. (1970). *Psi and Psychoanalysis.* New York: Grune and Stratton.

———. (1983). How to influence practically anybody (but fellow scientists) extrasensorily at a distance. In *Parapsychology and the Unconscious* (pp. 87–98). Berkeley, CA: North Atlantic Books.

Hastings, A., J. Fadiman, and J. S. Gordon (Eds.). (1980). *Health for the Whole Person: The Complete Guide to Holistic Medicine.* Boulder, CO: Westview Press.

Honorton, C. (1977). Psi and internal attention states. In B. Wolman (Ed.). *Handbook of Parapsychology* (pp. 435–472). New York: Van Nostrand Reinhold.

LeShan, L. (1974). *The Medium, the Mystic and the Physicist.* New York: Viking Press.

Mesmer, F. A. (1980). Dissertation by F. A. Mesmer, doctor of medicine, on his discoveries. In *Mesmerism: A Translation of the Original Medical and Scientific Writings of F. A. Mesmer, M.D.* (pp. 87–132). Trans. G. J. Bloch. Los Altos, CA: William Kaufman.

Myers, F. W. H. (1904). *Human Personality and Its Survival of Bodily Death.* London: Longmans, Green and Co.

Ostrander, S., and L. Schroeder (1970). *Psychic Discoveries behind the Iron Curtain.* Englewood Cliffs, NJ: Prentice-Hall.

Schlitz, M., and W. Braud (1997). Distant intentionality and healing: Assessing the evidence. *Alternative Therapies,* 3(6), 62–73.

Sicher, F., E. Targ, D. Moore, and H. S. Smith (1998). A randomized double-blind study of the effect of distant healing in a population with advanced AIDS. Report of a small scale study. *Western Journal of Medicine,* 169, 356–363.

Tart, C. (1969). Transpersonal potentialities of deep hypnosis. *Journal of Transpersonal Psychology,* 2(1), 27–40.

Vilenskaya, L. (1984). Psi research in the Soviet Union: Are they ahead of us? In R. Targ and K. Harary, *The Mind Race: Understanding and Using Psychic Abilities* (pp. 247–260). New York: Villard Books.

Winstead-Fry, P., and J. Kijek (1999). An integrative review and meta-analysis of Therapeutic Touch research. *Alternative Therapies.* 5(6), 59–67.

INTRODUCTION

In the early 1960s rumors began to circulate in the West that Soviet scientists had made decisive advances in the field of parapsychology, and the name mentioned as the top Soviet authority on the subject was that of L. L. Vasiliev. The Russian embassy in London distributed a leaflet containing his name in the context of parapsychological research.

At that time C. C. L. Gregory[1] and I were engaged on some research endeavoring to devise a theoretical framework that would, among other things, permit the occurrence of so-called paranormal phenomena. C. C. L. Gregory wrote to his colleague Professor V. A. Ambartsumian, whom he had met as a delegate of the International Astronomical Union, and inquired about Vasiliev's scientific and academic standing. The answer was reassuring: Leonid Leonidovich Vasiliev was Professor of Physiology at the Institute of Brain Research in the University of Leningrad and very highly regarded in his country as a scientist.

We then wrote to Professor Vasiliev asking his permission to translate and publish a work referred to in some of the publicity literature, *Mysterious Phenomena of the Human Psyche.* He replied in English that he would be delighted but he would much prefer us to publish his more scientific monograph, *Experiments in Mental Suggestion.* In due course two copies of his Russian paperback arrived. We arranged that Vasiliev would have the English version scrutinized and amended in Leningrad, and that we would accept his corrections as final (excepting matters of style).

We found two elderly Russian emigrés willing to make a literal translation of Vasiliev's book on the strict understanding that their names should never be mentioned. This version was literal indeed, since our translators were literary

---

[1]C. C. L. Gregory, my late husband, formerly director of the University of London Observatory and Head of the Department of Astronomy, who died in 1964.

rather than scientific, and quite unfamiliar with psychological and physiological terminology: one had to puzzle out, for example, that "absolute reflections" referred to "unconditioned reflexes." Also, our translators, who had obviously never seen a technical or scientific paper, were horrified and indignant at what they felt to be a barbaric insult to their beautiful language: they had to be assured over and over again that the style of an English scientific paper would be likely to differ as sharply from an English classical prose passage.

I taught myself the Russian alphabet and some basic rudiments by means of *The Penguin Russian Course,*[2] acquired Jablonski and Levine's *Russian-English Medical Dictionary,*[3] and proceeded to turn this translation into what I fondly hoped was moderately readable scientific English. The results were sent, chapter by chapter, to Vasiliev in Leningrad who amended and returned each part. As promised, we accepted his alterations as final. We printed the book ourselves by means of an old linotype caster and an ancient printing press; the binding was done professionally, as were the illustrations. It was published in 1963, without comment or introduction, as it stood, under the imprint of a small research organization, the Institute for the Study of Mental Images.

The book was, on the whole, very well received. To the best of my knowledge only one reviewer, Professor H. J. Eysenck, complained of its stylistic shortcomings, most of which were of course mine. There was no serious criticism of the work itself nor of the statistics. Even Professor Eysenck, whose review was probably the least favorable, stated that it was "obviously a 'must' for anyone interested in extrasensory perception," that so far as one could tell the experiments were "done under properly controlled conditions," and that "the statistical evaluation, although very simple, is essentially sound."[4]

As a matter of fact there were quite a few errors in the arithmetical calculations. We had intended to publish a revised edition soon after the first, which was sold out in a matter of weeks, rectifying the numerous typographical errors and correcting some of the tables with Professor Vasiliev's help. In order to rectify the tables one has to know whether slips occurred in copying, typesetting or proofreading on the one hand, or in the course of computation on the other. Unfortunately, C. C. L. Gregory died in 1964 and L. L. Vasiliev in 1966, and a second edition could not be published until now [1976].

On May 14, 1973, I wrote to Professor P. B. Terentieva, whose help with the statistics Vasiliev acknowledges in his preface, and sent her photocopies of the tables from the Russian version, and a list of the main ostensible errors. I

---

[2]J. L. I. Fennell (compiler), *The Penguin Russian Course, a Complete Course for Beginners* (Harmondsworth: Penguin, 1961).
[3]S. Jablonski and B. S. Levine, *Russian-English Medical Dictionary* (New York: Academic Press, 1958).
[4]H. J. Eysenck, "Soviet Experiments in Telepathy," *The Humanist,* 78, no. 12 (1963): 376–377.

asked her whether it would be possible to ascertain whether these were due to clerical slips or to errors in calculation. I pointed out that the errors were—if any—slight, and probably would not affect the argument or the conclusions, but it would be desirable to correct them and clarify the issues in a new edition. I received no reply. I have since learned that Professor Terentieva died some time ago.

I sent the same list of queries to Larissa Vilenskaya and to Eduard Naumov, both of whom I believed to be active in this field in the Soviet Union, with a similar request. Larissa Vilenskaya wrote to me on July 17, 1973, that she had forwarded the entire material to Dr. Pavel Gulyaev, who had worked with Professor Vasiliev. I have not heard from Dr. Gulyaev, and I received no reply from Eduard Naumov.

I reproduce below (p. xli) the list of figures I sent to Russia, and I acknowledge with gratitude Dr. P. Briskin's help in confirming and detecting numerical as well as other inaccuracies in the original text.

The reason why I received no reply from Eduard Naumov became clear in 1974. On Friday March 29, the *Neue Zürcher Zeitung* published a short paragraph to the effect that the Moscow parapsychologist Eduard Naumov had been sentenced to two years' hard labor. His alleged crime was to have misappropriated funds of a club where he had been lecturing on parapsychology. Dr. Andrei Snezhnevsky, whose name has repeatedly appeared as one testifying to the psychiatric instability of ideological dissidents, gave evidence for the prosecution. He did not, however, give it as his opinion that Naumov was in any way psychologically incapacitated; but he provided expert testimony that parapsychology is "a pseudo-science based on mysticism and idealism."

It would seem that the case against Naumov was entirely ideologically motivated. The two club officials who had, in fact, by all accounts collected the money for Naumov's lectures (which he was accused of taking) were declared mentally unstable and subjected to involuntary psychiatric treatment at the Serbsky Institute of Forensic Psychological Expertise, whose director is Dr. Andrei Snezhnevsky. No testimony on Naumov's behalf was admitted, despite the fact that a large number of witnesses came forward and offered to testify. *Samizdat* protested. Most of those in any way connected with parapsychology suffered serious professional difficulties. The ideological pendulum had once again swung against parapsychology.[5] Naumov, probably in response to an international outcry on his behalf, has been released from prison about a year

[5]L. Regelson, "An Appeal to Soviet and Foreign Public Opinion," *Samizdat,* translated June 25, 1974, by Caryl Emerson, published in the *Journal of the Society for Psychical Research* 47, no 762 (1974): 521–524; A. Gregory, "Is Russia Adopting a Party Line on Parapsychology?" *The London Times,* July 2, 1974.

before his sentence was due to end, but the intellectual climate in Russia towards parapsychologists would still seem to be hostile.

What exactly happened is not wholly clear but a coherent picture is slowly emerging. It came in the early 1960s as a very considerable surprise to almost everyone outside Russia that parapsychological investigations had been consistently and systematically pursued there under the most impeccable scientific auspices, the Institute of Brain Research of the University of Leningrad, by a former student and successor of academician V. M. Bekhterev.

It was Bekhterev who, in the early 1920s, had championed the cause of psychical research in Russia. Bekhterev was a reflexologist of international reputation, second only to that of I. P. Pavlov, and it is perhaps ironic to reflect that it was probably Bekhterev's influence,[6] more than any other, that led J. B. Watson to espouse and campaign for behaviorism, one of the most extreme reactions against a belief in the effectiveness of human conscious willing and imagining ever to gain widespread credence.

Bekhterev had, in the early 1920s, become interested in the claims of Vladimir Durov, a celebrated circus clown and dog trainer who entertained the public with a widely renowned "telepathy" act in which he "mentally" influenced his dogs to carry out numerous activities. Durov was an immensely successful man. He was a star performer before the Revolution, and our two translators instantly recognized his name, had seen him perform, and told us of the grand style in which he had lived and entertained in the old days. After the Revolution he became an established authority on animal behavior, partly because of his phenomenal practical experience with and understanding of animals; and partly no doubt due to his period of collaboration with Bekhterev.

Durov died in 1934, but his fame outlasted him: Emmanuel Dvinsky wrote a delightful and interesting as well as entertaining children's book called *Durov and His Performing Animals* in the late 1950s in which he describes some of Durov's activities.[7] Durov had one Alsatian dog called Mars, whom he not only trained to say "Mama" but apparently to respond to musical signals differing by quarter tones!

There was of course nothing telepathic about Durov's stage act which, like all regular, popular and public performances of this type, was based on well-rehearsed tricks. He trained and guided his dogs by means of a Galton whistle, which emits signals too high for most adult human ears but perfectly audible to dogs. (Since those days, supersonic signaling by means of a Galton

---

[6]G. Murphy and J. K. Kovach, *Historical Introduction to Modern Psychology* 6th ed. (London: Routledge and Kegan Paul, 1972), p. 246.

[7]E. Dvinsky, *Durov and His Performing Animals* (Moscow: Foreign Languages Publishing House, undated, some time between 1957—a date referred to on p. 117—and 1962, when I bought the book).

whistle has been suspected as the explanation of some of the late Dr. Soars telepathy with two young boys, though this has never actually been proved.[8])

Durov in the course of his circus activities became convinced that at times the impossible did happen: the dogs seemed to respond to what he merely thought. It was this conviction that Bekhterev investigated. Despite the very sensible precautionary measures taken by Bekhterev, such as the exclusion of Durov, the public, and Galton whistles, it is hard to be too impressed by the results obtained in these investigations. As Vasiliev puts it (see Chapter 4), it is not possible to accept a positive evaluation of these experiments without some reservations. It would seem that this line of investigation has not been pursued since the early 1920s. Nevertheless, Bekhterev considered that there was sufficient evidence to warrant further investigation.

In 1922 he founded the special Commission for the Study of Mental Suggestion attached to the Institute for Brain Research, specifically in order to investigate some of the phenomena of psychical research. Among its members were psychologists, medical hypnotists, physiologists, physicists and a philosopher. The activities of the Commission were extremely extensive and have, to the best of my knowledge, never been published in full. Among its activities was a substantial collection of case histories of people reporting psychical experiences, and the sifting of evidence and documentation related to such accounts. It would seem that the results obtained were by and large similar to those reported in the literature of the London Society for Psychical Research.

For example,[9] Boris Nikolaievitch Shaber, a student living in Witebsk, at 8:30 in the morning of December 17, 1918, had an hallucination of a luminous oval patch forming on his bedroom wall which grew and transformed itself into the image of his girl friend, Nadia Arkadievna Nevadovskaya, who was then in Petrograd. The vision smiled at Boris and uttered a phrase only the last word of which he was able to catch: ". . . *tlena*." The image darkened and disappeared, Boris Shaber wrote down an account of his strange experience the same day, and six people appended their signature to his account. On December 23, 1918, Boris received a letter from Nadia's mother informing him that her daughter had died at 8:25 on the morning of December 17, 1918. Her last words had been: "Boria (diminutive of Boris), there is neither spoiling nor decay" *("Boria, niet prakha niet tlena")*. Members of the Commission followed up the case, verified the signatures and satisfied themselves that the events described had indeed occurred.

---

[8]S. G. Soal and T. H. Bowden, *The Mind Readers* (London: Faber & Faber, 1959); see particularly C. Burt, "Experiments on Telepathy in Children—a Reply to Mr. Hansel's Criticisms," *British Journal of Statistical Psychology* XIII, pt. 2 (1960): 179–188.
[9]Cited by L. L. Vasiliev, in *La suggestion à distance* (Paris: Vigot Frères, 1963).

The Commission also studied the psychological and physiological effects of magnetic fields on human subjects. These results have not, as far as I know, been published.

Another of the Commission's investigations was concerned with distant mental suggestion of hypnotized human subjects. And it is this aspect of the work arising out of the Commission's activities that was extended and studied by Vasiliev who had recently joined the Institute for Brain Research as a young physiologist.

Vasiliev was disposed, on the strength of his own childhood experiences, to believe that there was something worth investigating regarding the phenomena of telepathy. The quotation introducing his popular book, translated into French under the title *La suggestion à distance* published just before the more detailed scientific book here introduced, is taken from the writings of C. E. Tsiolkovsky (1857–1935), a Russian inventor and pioneer of space flight:

> One cannot doubt the phenomena of telepathy. Not only is there a large accumulation of documents concerning these facts, but there does not exist a family whose members would refuse to testify to telepathic facts experienced by themselves. The attempt to explain these problems scientifically deserves our respect.

Whether or not there exist families whose members would decline to provide such testimony, in the Vasiliev family at any rate such experiences did pose problems. When Leonid Vasiliev was twelve years old and had just begun his second year of grammar school he, together with his brother and sister, went to stay with two young aunts in the family's country house near Pskov. His mother was at the time suffering from a serious liver complaint and had been taken to Carlsbad by his father. The Vasiliev children were enjoying a quite unaccustomed freedom and were one evening re-enacting the adventures of the children of Captain Grant, which involved the climbing of a sloping willow tree "to escape a flood." Leonid played the part of Paganel so thoroughly that, like his hero, he fell into the water. Unfortunately he could not swim and nearly drowned, and only managed to save his life by catching hold of the tip of a branch. His brother and sister watched the scene struck dumb with terror. All concerned seem to have been more frightened of the certain punishment awaiting them (especially as Leonid's much treasured and admired new white school cap was lost in the process) than of the very real danger of death. However, the young aunts agreed not to write to the parents about the incident, which, as Vasiliev dryly comments in a parenthesis, would hardly have been in their own interest, on condition that such activities should cease forthwith.

When the Vasiliev parents returned from Carlsbad the children and aunts were dismayed to be treated by Mama Vasiliev to a detailed account of the sorry tale, complete with willow, white cap, etc.: she had had a dream in Carlsbad of the entire incident, had awakened in tears and had insisted that Papa Vasiliev should instantly send home a telegram to make sure that all was well. Vasiliev, Sr. had decided to humor her, since, after all, she was ill. This consisted of getting up and dozing for half an hour in the entrance hall and subsequently assuring her (untruthfully) that he had wired home.

When Vasiliev joined the Institute for Brain Research in the early 1920s Bekhterev himself was conducting telepathy experiments, some of which were concerned with the distant influencing of dogs already mentioned, and some of which involved attempts to influence at a distance the behavior of hypnotized human subjects.

It is the topic of hypnosis that provides the background for Vasiliev's work, and it is useful to look at some of its antecedents. How does one individual influence another? Generally, this obviously happens in numerous ways, all of which involve some form of sensory communication. In the nineteenth century, when the phenomena of hypnotism created widespread controversy, the question was hotly debated whether the influence of a hypnotist over his subject was entirely due to what came to be known as "suggestion," or whether there was in addition some "fluidic" bond between the two.

The fluidic theory goes back at least to Mesmer who, towards the end of the eighteenth century, effected the most astonishing cures by means of an imposing mixture of ritual and magnets. Mesmer believed that his cures were due to, or mediated by, a special "fluid" filling the whole of space called "animal magnetism."[10] A French Royal Commission which subsequently investigated mesmeric cures carried out by Mesmer's disciple d'Eslon came to the conclusion that "the imagination is everything, magnetism nothing." In other words, the effects obtained were due to the impression made on the patient's mind by the doctor's activities as mediated by the patient's own ordinary sense organs. There was one dissenting voice: a biologist, L. de Jussieu, wrote a minority report of one, supporting the fluidic interpretation. He had observed some instances when in his view normal sensory communication could be ruled out, for example when a blind old lady responded appropriately to a "magnetized" rod pointed at her.[11]

The "suggestion" theory for explaining the overwhelming majority of hypnotic effects has gained almost universal acceptance. At first, in the course of the nineteenth century, the very phenomena of hypnosis were subject to just the type of skepticism that generally greets those of psychical research, and at least one professor of medicine lost his chair for claiming to perform major

---

[10] F. A. Mesmer, *Mémoire sur la decouverte du magnétisme animal* (Paris: Didot, 1779).
[11] L. de Jussieu, "Report of One of the Commissioners," Paris, 1784, quoted in R. Sudre, *Treatise on Parapsychology* (London: Allen & Unwin, 1960), pp. 19–20.

surgery without benefit of anaesthetics, by means of hypnosis alone. However, these phenomena are somewhat more amenable to experimental repetition, and, as scientists were increasingly satisfied that the feats of hypnotized subjects were not mere stage tricks, the notion of "suggestion" was endowed with ever greater explanatory power. As the astonishing plasticity and responsiveness of many ordinary people to hypnotic—and other than hypnotic—commands became accepted, "suggestion" came to be regarded as a suitable blanket term by means of which hypnotic and quasi-hypnotic effects could be explained. It seems to have widely escaped attention that "suggestion" is hardly much of an explanation, let alone a scientific theory.

No doubt one of the reasons why the term "suggestion" has been so widely adopted as an explanation is because it carries overtones of ordinariness and normality. We are, all of us, constantly subject to verbal and social influence, to wholesale suggestion, by those who constitute our "environment." We have a tendency to do as we are asked, to think in terms of the classifications and values embedded in our language and our culture, and to believe what we are told, however critical, rebellious and independent we may at times be.

Susceptibility to hypnotic commands may be viewed as merely a heightening of this normal tendency to think and to do as we are told, due to the artificial temporary abolition or attenuation of our independent personal ego in the hypnotic state. Thus, the wholesale employment of the concept of "suggestion" to describe hypnotic phenomena can be seen as a part of the modern process of "demystification" of the human scene.

There remain however, obstinate and venerable, the claims that the link between a hypnotist and his subject at times transcends the bounds of verbal suggestion; that one person can make another think and imagine and do things in ways not mediated by our ordinary organs of sense, possibly at a distance or across barriers sufficiently extensive to exclude normal sensory communication.

In order fully to appreciate the scientific background of Vasiliev's work, perhaps the most important historical reference is the International Congress of Experimental Psychology held in Paris at the École de Médecine in 1899.[12] The convener was the celebrated French hypnotist J.-M. Charcot, who was unable to attend himself owing to ill-health, his place being taken by Professor Ribot and Dr. Magnan as vice-presidents, and Professor Charles Richet (a Nobel Prize winner) as general secretary.

The Congress is significant for an understanding of Russian psychical research, for a number of reasons. For one thing, psychical research at that date was still an integral part of the study of physiology and psychology, and

---

[12]An account of some of the discussions and papers at the Congress is provided in A. T. Myers, M.D., *Proceedings of the Society for Psychical Research* VI (1889–90): 171–182.

questions concerning nonsensory communication were debated openly and on their merits. For another, the central topic of interest was the subject of hypnosis, its phenomena and explanations. Also the leader of the large Russian contingent was Professor B. J. Danilevsky from the University of Kharkov, to whom Vasiliev refers as one of the pioneers of the investigation of electrical processes in the brain, and who at that time was occupied in studying the hypnotizability of animals, including shrimp, crabs, lobsters, sepia, cod, brill, torpedo fish, tadpoles, frogs, lizards, crocodiles, snakes, tortoises, several species of bird, guinea pigs and rabbits.

Attention was inevitably focused on the rival theories of hypnosis: the "suggestion" and the "fluidic" interpretations. Professor Henry Sidgwick, the Cambridge philosopher, suggested three sets of condition under which suggestion was apparently excluded: in experiments (1) with animals, (2) with babies, and (3) at a distance.

Danilevsky thought that in animals the place of verbal suggestion was taken by physical manipulation of the animal, such as placing it in an abnormal position, enforced quiet, or gentle continuous pressure. Bernheim stated that similar considerations applied in the case of young babies: manual "passes" over the baby soothed him and bright lights tired his eyes, a sort of physical communication of suggestion could be adequately invoked as accounting for hypnotizability in babies, if indeed it was the same phenomenon.

There remained—and remains—Sidgwick's point about distant influence. Professor Delboeuf said that he himself had not been able to duplicate distant suggestion, while such results were indeed claimed to have been observed by many, including some of the most eminent scientists, scholars, and doctors attending the Congress, such as Janet, Richet, Ochorowicz, Sidgwick, and Myers. As Sidgwick pointed out, however, although hypnosis had been found to facilitate such effects, it was by no means indispensable.

Delboeuf's observation is of manifest importance here, in that it introduces the general problem of the unrepeatability of experiments in the field of parapsychology. If a man of the undoubted integrity, ability and open-mindedness of Professor Delboeuf is unable to duplicate the results obtained or witnessed by others of similar standing, does this imply that the latter savants must have been mistaken, either as regards their experimental procedures or else their interpretation of their observations? Presumably, one's answer here could be in the affirmative (and this was not a view adopted by Delboeuf) if and only if one is already convinced on quite other grounds that the effects are utterly impossible.

Such a dogmatic *a priori* negation of the very possibility of distant influence can indeed be found in the literature surrounding the field of psychical research. For example, Dr. G. Price, in 1955, took the uncompromising stand (which he subsequently retracted) that, since such linkage was miraculous and therefore impossible, he would elect to believe that scientists claiming such results were either mistaken or downright fraudulent. Price invoked in support the British empiricist philosopher David Hume who wrote that, if faced with the choice of believing in a miracle or else in man as a liar, he would prefer the latter. However, Hume is an unfortunate choice as a patron saint of dogmatic scepticism, since it was Hume's central tenet that we do not possess any *a priori* knowledge of the world—a view that implies that nothing is impossible, since we do not know what can and cannot be the case.[12]

Dr. Price was not alone in these views. In the 1956 edition of the *Soviet Encyclopaedia* the entry under "telepathy" called it "an anti-social, idealist fiction about man's supernatural power to perceive phenomena which, considering the time and the place, cannot be perceived."

Inevitably, if such a point of view is espoused, there is little point in experimentation: if results of distant influencing are claimed, the only thing to be found out is just *how* the error or the fraud came to be perpetrated, and psychical research amounts merely to a minor branch of, at best, psychopathology and, at worst, criminology. That psychical researchers of this persuasion are not lacking hardly needs elaboration.

But it is not an attitude that commended itself either to the participants at the Paris International Congress of Experimental Psychology in the late nineteenth century, or to the Russian investigators who carried out the work on distant influencing. Bernheim himself conceded at the time that if influence at a distance were indeed established, this would affect his suggestion theory and would, by implication, favor a fluidistic view, as championed by Professor Ochorowicz.

In Russia the use of hypnosis in psychiatric practice remained fashionable for longer than it did in the West, and it was, as Vasiliev describes, some demonstrations by Dr. K. I. Platonov at the 1924 Second All-Russian Congress of Psychoneurologists, Psychologists and Teachers that revived interest in distant influence.

As Platonov himself describes (see Appendix E) he had come prepared to read a paper on his experiments in distant influence without any intention of providing an experimental demonstration. On his arrival in Leningrad he accidentally met one of his ex-patients and subjects, M., in the street and asked her to accompany him to the Congress, to which she agreed without any idea that she was to be the subject of an experiment.

The demonstration was sufficiently impressive for the Congress to include among its resolutions a recommendation to the effect that the phenomena of distant suggestion were worthy of further scientific study, and entrusted their further exploration to the Society for Neurology, Reflexology, Hypnotism and

[13]A. Kohsen (A. Gregory), "Science and the Supernatural," *Journal of the Society for Psychical Research*, 1956, 38, no. 687 (1956): 226–227.

Biophysics attached to the Institute for Brain Research. Vasiliev was put in charge of one of the Society's sections.

Platonov, like Vasiliev, had been a former student of Bekhterev's, under whose direction he had originally completed his post-graduate research on the mechanisms of verbal suggestion in hypnotic sleep. Platonov was a professor at the University of Kharkov, one of Danilevsky's successors. His position in Russia became one of great eminence and his book, *The Word as a Physiological and Therapeutic Factor,* which runs into over 450 pages, remains an authoritative work on medical hypnotism. A revised edition appeared in an English translation in 1959.[14]

Whatever one might think about some of Platonov's theoretical reasoning, the wealth of experimental and clinical material and its practical importance is overwhelming. Of special interest are the experiments in which physical symptoms, such as vomiting in response to massive injections of apomorphine, could be completely inhibited by means of hypnotic suggestion. There is also an account of how Dr. V. Finne, incidentally one of Vasiliev's most effective hypnotists at a distance, demonstrated to Professors Chernorutsky, Povarnin, Platonov, and others, the creation of heavy burns in response to mere suggestion, and there is a photograph of a second-degree burn produced in this manner. Similar experiments were repeated later, when frostbite, rashes, and "pigmentation" were added to the list of physical symptoms that could be produced by suggestion. A drop in blood sugar level was demonstrated when it was hypnotically suggested to a subject that he was drinking distilled water, when in fact he was fed a concentrated sugar solution (which would, of course, without contrary suggestion, send the blood sugar level soaring).

The sheer scientific interest of Platonov's work should not be allowed to eclipse the astonishing therapeutic achievements documented in his book. Among the complaints to have been completely cured in a few hours of hypnotic therapy (none more than twelve hours) were: phobias, hysterical complaints, depressions, diabetes *insipidus* and *mellitus,* toxaemia in pregnancy, baldness, weeping excema (in one case of fourteen years' standing), warts, vomiting in pregnancy, paranoia, obsessional syndromes, impotence, the chronic phase of traumatic neurosis (notoriously recalcitrant to psychotherapy), epilepsy (including *grand mal*), tics, allergies, neuralgia, alcoholism, morphine and cocaine addiction, and hyperthyroidism. Follow-up studies revealed that the patients remained in good health with full capacity to work for many years.

Professor Platonov's standing and importance in the Soviet Union is not surprising in the light of his prodigious successes in the fields of experimental

---

[14]K. I. Platonov, *The Word as a Physiological and Therapeutic Factor* (Moscow: Foreign Languages Publishing House, 1959).

physiology and therapeutic medicine, and one can well understand Vasiliev's delight when, more than thirty years after the demonstration on M., of hypnosis at a distance, Platonov sent him an account of his own recollections of the experiments (see Appendix E), as well as letters concerning some further and independent experiments on distant influencing carried out at the University of Kharkov by Professor Dzelikhovsky and Drs. Kotkov and Normark. The observations, sketchy though they are, closely resemble those of the French hypnotists and the many instances of distance influencing recorded in the literature of psychical research.

If communication can be established between two people, does there have to be a sensory link between the two? If communication is believed to have occurred, and a normal sensory mode of perception is thought to be excluded, does it follow that "non-" or "extra-sensory" perception or communication has been established, i.e., nonphysical communication? Or does it follow that there must be some physical nexus or channel that has to be discovered? Stated thus baldly, the dichotomy sounds crude: after all, even ordinary sense perception is universally believed to have a reasonably well-establsbed physical nexus in terms of, say, light and sound waves, and people do not therefore necessarily believe that all sensory perception is "merely physical."

However, this is part of the theoretical battle that underlies a great number of the controversies surrounding parapsychological investigations. It is widely believed by those championing parapsychology that a proof of extra-sensory perception or influencing would prove a nonmaterialistic theory of the universe, in that here there would be demonstrable events without a physical basis. This assumption seems to me to be mistaken. It is not possible in the present context to go in any detail into variants of materialism, idealism, dualism, and so forth, beyond a few remarks that have a direct relevance to the work of Vasiliev.

Let us suppose that extra-sensory mental suggestion at a distance has been established as having taken place. How would a scientist go about further exploring this phenomenon? He would presumably first of all attempt to test for a particular physical nexus which is known to mediate communication between humans at a distance; and a very obvious candidate for such a channel is communication by means of radio waves.

The problem as it presented itself to Vasiliev in the early 1920s was that, once the phenomena of distant influence could be reproduced in the laboratory with reasonable reliability, the next and obvious step was to investigate their physical basis. If messages travel from A to B they must traverse space, and some physical nexus carrying the transmitted information must be detectable. Both Bekhterev and Vasiliev believed that the answer to this problem had been supplied by the Italian neurologist Cazzamalli, who had published several papers championing the electromagnetic theory of telepathy, which is of course a variation of Mesmer's fluidic theory.

In Cazzamalli's view the information conveyed from sender to recipient was carried by electro-magnetic energy in the form of radio waves ranging from 0.7 to 100 meters. In 1926 Vasiliev published in the Russian journal *Science News* a paper entitled "The Biophysical Foundations of Direct Thought Transmission," in which, as he says, he expounded "a materialistic approach to the phenomena of mental suggestion."

The bulk of Vasiliev's work after this date was devoted to an attempt to establish Cazzamalli's radio-brain-wave theory. The book here introduced traces in detail how, after years of ingenious, systematic and painstaking study, the Russian team, to their surprise, found that Cazzamalli's theory was in fact incompatible with their observations: metallic barriers, which would stop all radio waves of the requisite frequencies, completely failed to screen out the direct mental influence of the hypnotist on his subject. Why was the work suddenly discontinued in the mid-1930s, only to be resurrected some thirty years later?

It is impossible to begin to answer this question without some considerations of the political and ideological controversies surrounding parapsychology. As has been indicated, parapsychological experimentation is widely (though in my view erroneously) believed to be crucial to a decision between a "materialist" and a "spiritual" view of man. Traditionally parapsychological claims are associated with religious beliefs and mystical experiences, and no government officially opposed to religious beliefs could be entirely neutral about a set of experiments believed to have a direct bearing on such beliefs. And indeed, parapsychological observations might well have a bearing on particular beliefs, even if the data are, as I have suggested, in principle quite unsuited for discriminating between rival metaphysical systems.

We have already seen that in the early days of the Soviet Union there was no official opposition to psychical research. On the contrary, as Vasiliev mentions in his preface, in 1924 Mr. A. V. Lunakharsky, Commissar for Education, himself took the initiative in forming a Soviet Committee for Psychical Research to be affiliated to the International Committee for Psychical Research just founded by Carl Vett on the initiative of Charles Richet.

Now as long as Vasiliev believed that he was substantiating Cazzamalli's brain wave theory of telepathy, he was secure from any reproach of anti-materialist heresy, and he was manifestly encouraged and financed in the usual way. When it became ever clearer that his results did not support the brain wave theory, funds and encouragement withered away. The definition in the *Soviet Encyclopaedia* for 1956 of telepathy as anti-social and impossible considering the time and the place is uncompromising enough. During the later 1930s, '40s and '50s all periodicals and papers dealing with psychical research sent to Russia from abroad were "returned to sender." In Professor Platonov's magnum opus, *The Word as a Physiological and Therapeutic Factor,* there is not one hint, from cover to cover, that the author had been a champion of telepathic influencing, and continued to remain convinced, as emerges clearly from the appendices to Vasiliev's book.

In 1959 there appeared in the French popular journal *Constellation* an article by Jacques Bergier entitled "La Transmission de Pensée, Arme de Guerre" ("Thought Transmission, a Weapon of War"), followed by another, early in 1960, by Georges Messadié in *Science et Vie,* called "Du *Nautilus*" ("On the *Nautilus*"). These articles described telepathy experiments alleged to have been carried out aboard the U.S. nuclear submarine *Nautilus.* According to these stories, shore-to-ship thought transmission of information had been entirely successful.

Radio communication between submarines and the outside world constitutes a notorious military problem, since the combination of a thick layer of sea water and the hermetically sealed metallic covering of a submarine effectively screens out or seriously attenuates radio waves. Are there telepathic methods of communication capable of piercing barriers of sea and metal?

The truth or falsehood of the claims for the *Nautilus* experiments is still shrouded in mystery. Dr. J. B. Rhine stated that "authoritative sources in Washington denied all knowledge of such experiments" (which, unfortunately, one might expect authoritative sources to do whether or not such experiments had in fact taken place).

From the Russian point of view, however, such claims were of the utmost importance, not only for strategic reasons too obvious even to mention, but also because Vasiliev, Professor of Physiology in the University of Leningrad and by now a corresponding member of the Soviet Academy of Medical Sciences and a holder of the Order of Lenin, had already conducted experiments in mental influence that had apparently conclusively shown that (a) distant influence can be experimentally demonstrated and (b) that such remote influence was not affected by metallic screening of just the type to exclude radio signals. However, as already mentioned, his work had been discontinued because of the official Marxist party line on telepathy at the time, although his records had been preserved in the archives of Bekhterev's Institute for Brain Research in Leningrad. Vasiliev, like Platonov, is careful to maintain that we are dealing here with a phenomenon which, however important its scientific implications, has nothing at all to do with "idealism" or religion: doubtless it would only be a matter of time before these observations could be shown to have a "materialistic" basis just like other facets of nature.

Professor A. V. Tugarinov, head of the Department of Philosophy of the University of Leningrad, declared in favor of research in telepathy: in the *Leningrad University Messenger,* 1964, he wrote that preconceived ideas and ideological processes presented the most formidable obstacles to objective research. Rather than engage in theoretical arguments, scientists should seek to explain the energetic and physiological basis of such phenomena. Tugarinov went further: "All critics of telepathy are only using Marxism-Leninism to support their scientific conservatism. All who throw obstacles in the path of scientific progress should be made to suffer."

Not all of us are quite so uninhibited in expressing our sentiments about the fate we wish to befall those who are so reprehensible as to disagree with us. But the gist of all this argumentation is clear, and not at all dissimilar from the type of controversy that has always surrounded psychical research in the West. Do the phenomena of distant influence, if they happen, undermine a scientific view of the world? Do they support a more mystical or spiritual or even religious interpretation of the universe? Can they, in principle, be explained in terms of present-day physics?

The term "mystical" is a term of abuse in both East and West, and would seem to require considerable analysis and clarification before it fulfills any function more useful than that of serving as an indication of disapproval. In Russian writings the word "idealistic" is often employed to contrast a concept with an acceptable "materialistic" one.

Vasiliev (see Chapter 10) castigates those who would oppose telepathy research on scientific grounds as being "mechanistic materialists." Normally, of course, Marxists contrast "mechanistic materialism"—which they reject and which they deem typical of Western ideology—with "dialectical materialism," to which they themselves subscribe. Vasiliev, on the other hand, contrasts "mechanistic materialism" with an acceptance of cybernetics and, by implication, with information theory which he considers to be "materialistic" and not necessarily "mechanistic."

To the extent to which, in explanations involving communication or information theory, the emphasis is shifted from the energetic manner of communication to the meaning of the message conveyed, Vasiliev is in a reasonably strong position. On the other hand, to the extent to which the particular cybernetic model he cites—that of M. Raphael Kherumian—is plainly unsatisfactory even to Vasiliev himself, and since no physical channel across which the information could flow has been discovered to date, Vasiliev is on much weaker ground.

As he says, abroad (that is, in the West) there is no shortage of supporters of the "'psychic' not to say frankly 'spiritualistic' hypothesis which separates the psyche from the brain." A great many parapsychologists have indeed taken the line that, if the psychical phenomena could be firmly established, this would prove the independent existence of "the mind" as contrasted with "matter," though most parapsychologists have also been at pains to point to a fundamental barrier separating "psychic" phenomena such as telepathy from "spiritualistic" beliefs in the continued existence of the soul after death.

Actually the argumentation surrounding this whole topic has, to date, spread more heat than light. After all, supposing a physical basis for thought transmission were to be established, this could be interpreted (depending on just what was found) as demonstrating a possible "materialistic" basis for consciousness to function separately from the anatomical body and brain.

It seems to me extraordinarily doubtful whether the demonstration of any phenomena whatsoever could, in principle, disprove (let alone prove) fundamental metaphysical positions such as materialism (mechanistic or dialectical variety) or substantive dualism (whether psychological or spiritualistic) or for that matter pure idealism, which no one at the moment appears to champion.

On the other hand, the difference made to our imaginative picture of man and the universe, if the phenomena of psychical research were to be systematically admitted is, I believe, immense. This has nothing to do with their supposedly "nonphysical" characteristics. It seems to me highly probable that, once physicists and mathematicians begin to construct experiments and theoretical models to clarify the phenomena of distant influence they will also devise ways of giving lawful and quantitative accounts of these happenings, as well as discovering physically detectable concomitants. It is not as if physics had, even in principle, explained everything that occurs on the surface of our planet, barring only telepathy!

The problem does not seem to be one of *physical grounding* but of *anatomical location.* Vasiliev states categorically that our psyches must never be separated from our brains. This is, surely, simply a confession of faith. There is no logical necessity for this to be so, however plausible the assertion might be. Hence it would seem to be a matter for factual investigation and not for unqualified axiomatic assertion. As has been argued by Professor P. F. Strawson, the fact that for each person there is one particular body that occupies a highly privileged relationship between experience on the one hand and anatomical position on the other, is contingent and not logically necessary. ("Contingent" in philosophical argument means that something happens to be so, but might be otherwise. Empirical investigation is concerned with how the world happens to be.) Strawson[15] posits a logically possible state of affairs in which a subject of experience, S, has three bodies, A, B and C, supplying S with different facets of visual experience. This is, of course, fanciful and Strawson does not suggest that he or anyone else actually believes such a state of affairs to obtain anywhere: he merely uses it to illustrate the difference between what is the case, what could be the case, and what must be the case.

Now some of the phenomena of distant influence seem actually to point to a state of affairs not so very remote from Strawson's flight of the philosophical imagination. For example, the German doctor Gustav Pagenstecher describes experiments with a subject, Maria Reyes de Z., who, once she was in hypnotic trance, lost all powers of sensory experience: she could neither see nor feel nor smell nor taste nor hear, except that she continued to hear and respond

---

[15]P. F. Strawson, *Individuals* (New York: Barnes & Noble, 1965), p. 90 ff.

to the hypnotist's voice. However, she did have the same sensory experiences as Dr. Pagenstecher had undergone: if he put salt or sugar on his tongue she would experience the taste of these substances at that time; she heard the ticking of a watch held to his ear; she blinked when his eyes were subjected to a flash of light; she sneezed when an ammonia flask was held under his nose, she experienced a needle prick in his finger in precisely the same corresponding location of her own finger.[16] Perhaps the world *is* at times such that what is experienced by one body is experienced by another subject located elsewhere. Pagenstecher's is not an entirely isolated case: the German psychical researcher Tischner made some similar observations, and I have seen recently, at an international Congress in Prague (June 1973), a film made by Dr. Z. Rejdak in which apparently subjects under hypnosis experienced the effects of physical and physiological stimulation applied to another person whom they were unable to see. Russell Targ and Hal Puthoff describe how subjecting a person to certain visual experiences can apparently affect the alpha rhythm of the brain of another person located elsewhere—a related and, if confirmed, exceedingly important observation.[17]

Vasiliev, as well as Platonov and many others, found that success at distant influence depended among other things upon the hypnotist's vividly imagining that the subject was in fact doing or experiencing what he wanted him or her to do or experience. Whatever the ultimate physical nexus between the location of the bodies of hypnotist and subject, what was vividly imagined by the experimenter in place p_1 was experienced by the subject at the same time in place p_2. And this throws considerable doubt on the total insulation in principle of separate experiencers anatomically located in different places. It suggests that, as a matter of contingent fact, a person can "perceive" and experience events to which he does not have relevant anatomical access. In other words, under certain conditions people can "perceive" and respond to stimuli to which their bodies have no access *via* their normal sensory equipment.

Once the possibility of distant influence is granted, the door is opened to (among other things) entirely novel ways of explaining cultural, social, personal and physiological similarities and differences between persons, or for that matter individual members of any species. It could, for example, throw a completely new light on current controversies concerning language acquisition: Professor Noam Chomsky believes that, in order to understand the rapidity and the logical processes by which children learn to speak, we require theoretical constructs such as "deep structure" which must be inherited. This has

---

[16]Cited in W. H. C. Tenhaeff, *Aussergewöhnliche Heilkräfte* (Olton: Walter, 1957) p. 104 ff.
[17]R. Targ and H. Puthoff, "Information Transmission under Conditions of Sensory Shielding," *Nature*, 251 (Oct. 18, 1974): 602–607.

involved him in postulating the genetic transmission of innate ideas.[18] If distant group telepathic influencing were allowed, this might lead to new theories and new empirical and experimental observations in the immensely important field of language learning.

Closely beneath the surface of controversies ranging around the phenomena of psychical research both in the West and the East is the ever-present specter of religious beliefs, particularly in a continued existence after death. For Marxists, such beliefs seem particularly pernicious from a social point of view in that it might quite reasonably be expected to detract from our efforts to make this planet here and now into a somewhat less unsatisfactory place; but Western champions of present-day science as a system of belief are hardly any less zealous in denying as mystical, obscurantist and absurd the very possibility of human survival of death. Vasiliev is at special pains to dissociate himself from spiritualistic beliefs, just as many Western exponents of parapsychology have been eager to affirm that psychical phenomena have little if anything to do with beliefs in survival. Such a strict division seems to me open to debate if the tie between anatomical location and experience is loosened. Once it is supposed that our veridical experience (that is, experience of what is in fact the case) is not fully dependent on what reaches our brain via our organs of sense, then there is some degree of independence of person from anatomical body. The question again becomes one of degree, not of kind—of "how much?" rather than of "whether?".

Professor Strawson sees no difficulty in our intelligibly envisaging our own personal survival of bodily death, for the simple reason that we can quite easily imagine it. He merely imposes two limiting conditions: that (a) one should have no perception of a body related to one's experience as one's own body; (b) one should have no power to initiate changes in the physical conditions of the world such as one would normally be able to make with, say, one's hands and vocal chords. If I understand him correctly, Professor Strawson imposes these conditions on the grounds that not to impose them would be a "vulgar fancy" and would take one into the realm of spiritual seances.

Now it seems to me that vulgarity is neither here nor there, and that the dividing line between the philosophical imagination and vulgar fancy is rather a blurred one. That some contingency is vulgar is hardly a good reason for saying it could not be true; nature is no great respecter of proprieties. Perhaps Professor Bernard Williams[19] is right when he states that the imagination is too tricky a thing to provide a reliable road to a comprehension of what is logically

---

[18]N. Chomsky, "Recent Contributions to the Theory of Innate Ideas," *Synthese* 17 (1976): 2 ff.

[19]B. Williams, *Imagination and the Self*, Annual Philosophical Lecture of the British Academy, 1966.

possible; at any rate, imagination and imaginability are tricky guides for ruling out what can*not* be. The formulas of modern relativity theory and quantum physics describe states of affairs which it is not possible (at any rate, for most of us) to imagine. In fact, mathematical language has been devised precisely in order to enable us to go beyond what can be said in ordinary speech or conceived in terms of everyday imagination. But this does not prevent me from accepting, perhaps merely as a characteristically pious child of the twentieth century, that they describe important aspects of scientific reality. And, after all, if Jean Piaget has shown anything, he has established that some children at certain stages of development are at times quite unable to imagine the world from other than a severely restricted egocentric one—which does not prevent these other perspectives from holding true for other experiencers!

When Kitaygorodsky and Roshchin were arguing about the impact of an acceptance of telepathy upon people's religious beliefs, they were agreed that a belief in human personal survival is both false and pernicious: only Kitaygorodsky was (at least before his conversion) convinced that a study of telepathy would favor such beliefs; whereas Roshchin contended that a scientific study would exorcise the spirits. Time will show which, if either, of them will turn out to be right.

Vasiliev certainly had the good sense to steer clear of this particular issue: there is only one single instance which might conceivably have a bearing on relevant mediumistic phenomena. In Chapter 5, in Task 5 given to his student C., Vasiliev tried to transmit the name of a dead girl. C. "automatically" wrote the name of another girl, Tamara, whom she disliked, together with scribbles suggesting death. She then went into trance. Vasiliev woke her up very sharply, and she declined to cooperate further. Vasiliev gives what most psychoanalysts would accept as a reasonable interpretation, namely that C. wanted Tamara to die, and her refusal to cooperate further might be said to have constituted a defensive reaction against death wishes towards the living Tamara. Whether a given analyst would accept Vasiliev's further supposition that there was some element of telepathic triggering off (by means of the transmission of a female name coupled with the idea of death) depends on whether the particular analyst would accept the notion of telepathy. Freud himself did, and it seems to me just the sort of interpretation he might well have given of the incident.

On the other hand, one cannot help wondering why Vasiliev, who certainly lacked neither patience nor curiosity, should have so abruptly and harshly terminated the experiment. It occurs to me that a mediumistic seance in which C. conjured the image of the dead girl whose name Vasiliev had in mind, could have proved (to put it mildly) embarrassing.

Actually, as anyone who has ever seriously engaged in psychical research is only too well aware, investigations of mediumistic phenomena present problems so intricate and difficult that we are not yet in a position to do much more than tell the stories and wonder what to make of them—a thoroughly unsatisfactory state of affairs for any systematic, let alone scientific, study. However, it seems relevant to say something about the issues involved, since fundamental ideological considerations bearing on the interpretation of psychical research data are drawn into question. It might be helpful to give an actual instance to illustrate some of the difficulties. The case is quoted by Alexander N. Aksakov, a nineteenth-century Russian psychical researcher[20] and is one which F. W. H. Myers thought constituted particularly good evidence for human personal survival.

A Russian lady, Madame von Wiesler, and her daughter, Sophie, held some seances in their home, using a saucer with a black line and a homemade alphabet. One day, on January 22, 1885, the saucer spelled out the name "Shura" and announced by pointing to the appropriate letters that it was up to Sophie to save Nicolas.

The ladies asked what this meant, and the reply was spelled out that Nicolas was compromised, just like Michael; Sophie must immediately go to the Technological Institute, ask for Nicolas and make an appointment with him. He was being led astray by a set of good-for-nothings and must be warned. The ladies were not to bother about silly social niceties.

Now "Shura" was the name of a young girl of seventeen who had killed herself by swallowing poison about a week before the seance. She had two cousins, Michael and Nicolas. Michael had become involved in revolutionary activities and had been arrested, imprisoned and killed while attempting to escape. His cousin Shura had shared his revolutionary sympathies. These facts were of course known to all in St. Petersburg.

Mrs. von Wiesler and her daughter now found themselves in a quandary. They knew quite enough about seances to be aware of the unreliability of such "communications," and all the social proprieties made it extremely difficult for Sophie to go to the college of the young man, who was merely a distant acquaintance, and tell him that the ghost of Shura had sent her to warn him! They decided to do nothing, though the saucer's messages continued their urgent entreaty. Eventually Mrs. von Wiesler capitulated to the extent of telling Nicolas' parents in the strictest confidence what had happened, but they were entirely satisfied that there was no truth in the supposition that their other son was involved in any revolutionary activities. Nothing happened during the next two years, and mother and daughter became convinced that it had all been a lot of embarrassing and pernicious nonsense.

On March 9, 1887, the Tsarist secret police suddenly raided Nicolas' rooms, arrested him, and he was exiled from St. Petersburg. It emerged that he had taken part in anarchist assemblies which were held during January and February, 1885. There had been no close connection between Sophie and

---

[20]A. N. Aksakov, *Psychische Studien* (Leipzig, 1889).

Shura, merely a few short meetings when they had been at school together at a time when Sophie was thirteen and Shura still younger.

Now, so far from affording peculiarly satisfactory proof of personal survival, the case seems to me to lend itself to a number of different interpretations in addition to the "vulgar" one (to use Professor Strawson's term) that the departed Shura was trying to save Nicolas. For instance, it does not seem beyond the bounds of possibility that Sophie, like Nicolas and Shura, had been involved with anarchist activities, had taken fright, and was using the seances to get her mother to extricate Nicolas from his dangerous entanglements. True, Sophie never seems to have put the slightest pressure on her mother to pursue any contact with Nicolas, but in the circumstances this could have given the game away.

Again, Sophie might perhaps have wanted an excuse to meet Nicolas and either deliberately or unwittingly pushed the saucer. The dates and the subsequent arrest would in that case be coincidences, evidently not so very farfetched in the political climate of Russia in the mid-1880s.

Supposing these two entirely "normal" explanations were ruled out, but telepathy were allowed, Sophie or her mother could have had some extrasensory information concerning Nicolas' activities, and unconsciously dramatized this information, putting the warnings at the door of Shura, whose tragic suicide only a week before must surely have exercised people's minds in St. Petersburg society at the time.

It is simply not possible, on the strength of such evidence as we have to date, to come to any confident conclusion about the way or ways to interpret these mediumistic phenomena, which have not changed much since Aksakov's day. But that they are by no means deeply buried beneath the surface of the minds of those who concern themselves with psychical research, in the West or in the East, is quite certain. It seems most unlikely to me that Vasiliev was ignorant of the extensive and detailed writing of Aksakov.

Quite apart from the politically and ideologically tricky issue of personal survival, ostensible telepathic experiences do tend to present a problem to those who have them and who are committed to a picture of the world that would exclude them. Vasiliev was permitted in 1960 to establish a unit at the Institute for Brain Research that concerned itself specifically with parapsychological matters. As soon as this became known, he was inundated from all parts of Russia with letters from correspondents giving details of psychical experiences which had troubled them deeply. He cites a particularly poignant instance given by a teacher, Madama Agenossova, living in Nishni Tagil,[21] who says how ashamed she had been all her life of having telepathic experiences, since this is so unbecoming to a good Communist. Throughout her life she had had dreams and visions which reliably supplied her with information she ought not to have had.

---

[21]L. L. Vasiliev, *La suggestion à distance* (Paris: Vigot, 1963): 27 ff, p. 86 ff.

For example, in 1942 she suddenly had a vision of getting a telegram from her husband stating that he was being sent to the front from Sverdlovsk. Actually be was stationed in Shadrinsk and was not due to depart until later. She was so certain that the vision was right that she instantly obtained the necessary traveling papers to go to Sverdlovsk. There she went to see her daughter, who was married to a member of the N.K.V.D., and, while they were talking, Mrs. Agenossova's husband turned up and said that he had written out just such a telegram but had decided not to send it in order not to upset her. The incident was witnessed and testified to by her son, V. V. Agenossov, headmaster of Secondary School No. 32.

She also dreamed during the war that her husband and her son, who had been posted to different parts of the front, had met. She wrote to both of them at once and found that on the day of her dream they really had met, a totally unexpected and highly improbable event. She states that later she dreamed that her son had suffered a concussion and he had; that her husband had had an affair with another woman and he had. She had never dared to speak of these things outside her family circle, it was all so un-Communistic and unscientific, but now that scholars and scientists were concerning themselves with such matters perhaps it was all right, and it might even perhaps be helpful to write about them.

As has been shown, the political and ideological acceptability of parapsychological research in the Soviet Union has been subject to considerable fluctuations: flourishing openly at first under Bekhterev and Platonov in the 1920s; then, at best discouraged, at worst suppressed when Vasiliev's simple materialist theory had not been proved in the late '30s, '40s and '50s; acclaimed and financed in the early 1960s when there seemed to be some risk that American scientists might claim the credit for discoveries priority for which quite properly belonged to Russian scientists; and persecuted once again in the late 1960s and 1970s.

What brought about the renewed change of policy after Vasiliev's death?

Dr. J. G. Pratt, one of the first Western researchers to visit Russia after the publication of Vasiliev's work, describes two of his visits, one in 1963, the other in 1968. He provides a vivid demonstration of the difference in atmosphere pervading the two conferences he attended, both of which were organized by Eduard Naumov. During the former, free and cordial exchange of views was possible; the second was more or less wrecked by an article in *Pravda* attacking parapsychology. Most of the Russians declined to deliver their papers and Western visitors were pressed to give impromptu lectures; the House of Friendship withdrew its invitation to hold further meetings or allow films to be shown.[22]

---

[22]J. G. Pratt, *ESP Research Today* (Metuchen, N.Y.: Scarecrow Press, 1973), pp. 55–83.

From this time onwards official hostility towards parapsychology increased in the Soviet Union. What seems to have happened is that the Russian authorities took the strongest possible exception to a book by two Canadian journalists, Lynne Schroeder and Sheila Ostrander, *Psychic Discoveries behind the Iron Curtain,* based on a visit to Russia and other Eastern bloc countries by the authors in 1968.[23] The "Voice of America" beamed a radio programme into Russia discussing the Schroeder and Ostrander visit, and it would appear that the tenor of this broadcast was such that the two journalists' visit and their subsequently published book could be construed as a politically motivated attack using parapsychology as a propaganda weapon.

Unfortunately for him, Eduard Naumov had acted as guide, mentor and interpreter to the two ladies during their stay in Russia, and his help and advice is acknowledged throughout the book. Apart from the "Voice of America" episode it is not clear why Soviet officialdom should have taken such fierce exception to a candidly popular and sensationalistic book, which is not at all of a type likely to be taken particularly seriously by Western scientists. Perhaps the most plausible interpretation of the Russian reaction is that they are quite understandably worried that they might be believed to be, by the world's scientific community, self-proclaimed champions and leaders of parapsychology, especially as expounded by Schroeder and Ostrander. So far as officially sanctioned Soviet science is concerned, this is simply not true. Russian scientists are just as divided among themselves concerning parapsychology as are scientists in the West.

In October 1973 four eminent members of the Moscow Academy of Pedagogical Sciences—Professors Zinchenko, Leontiev, Lomov and Luria—wrote a long and detailed paper entitled "Parapsychology—Fiction or Reality?" This was published in *Questions of Philosophy,* an official publication of the Soviet Academy of Pedagogical Sciences, and explicitly sets out "to express the viewpoint of the USSR Society of Psychologists towards parapsychology."[24]

The paper is certainly very interesting. The authors accept that "obviously some so-called parapsychological phenomena do happen; however, the main obstacle to the acceptance of their existence is ignorance of the basis of their operation." It is by no means wholly clear from their paper just which of the parapsychological phenomena the authors believe "obviously happen," since the only ones they unambiguously support as authentic, such as Kirlian photography and Rosa Kuleshova's "skin vision," are explicitly stated not to be

---

[23]L. Schroeder and S. Ostrander, *Psychic Discoveries behind the Iron Curtain* (New York: Prentice-Hall, 1970).

[24]V. P. Zinchenko, A. N. Leontiev, B. F. Lomov, A. R. Luria, "Parapsychology—Fiction or Reality?" *Questions of Philosophy* 27 (1973): 128–136. Translated by M. Mihialovivusic and C. Bird, edited by S. Krippner (24 pp.).

parapsychological after all. Indeed, Rosa's "dermal-optical vision" is said to have "nothing in common" with parapsychological phenomena: "it would seem that the dermal phenomenon actually exists . . ." Zinchenko *et al* accept a definition of parapsychology as studying, among other things, "forms of perception affording a means of receiving information which cannot be explained by the known senses"; and it is not easy to see why, if this is so, Rosa's apparent ability to perceive via her skin impressions normally requiring eyes, should fall outside the realm of parapsychology.

Perhaps phenomena are being tacitly classified into two categories: those that happen (and hence are not parapsychological) and those that do not (and hence are). The authors are indeed at pains to show that there is no such subject as parapsychology, only motley phenomena that belong into various other academic disciplines. A detailed discussion of this view, for which a case can indeed be made out, is out of place here. However, the political implications are relevant. A large proportion of the paper is, in fact, devoted to denunciation of "militant parapsychologists" and, although just who is being attacked is not made fully explicit, the implications are reasonably clear: "Militant parapsychologists" are those who have championed the subject in Russia since Vasiliev's death—notably Eduard Naumov. He is not mentioned by name, but the Institute for Technical Parapsychology is, as an instance of the type of organization which does not, and never has, existed in the USSR. The nonexistent Institute for Technical Parapsychology, as described by Stanley Krippner, was the presumably private organization of which Naumov was director.[25]

Zinchenko and his colleagues state that one encounters in the literature speculations on the theme of national defense, psychological warfare, intelligence gathering, etc., and they shrewdly surmise that such speculations are apt to be designed to obtain government finance for parapsychological research. "However, the general stream of parapsychological literature simply does not contain evidence that these applications of parapsychology are frequently made." Yet later on in the article the authors write that "Parapsychologists often have an entirely practical motivation, for instance to discover through their study of telepathy a new means of communication in order to transmit important information, or through their studies of psychokinesis to discover a new form of energy in order to trip the detonating switch of an instrument at a distance." One can hardly imagine any defense department anywhere in the world entirely disinterested in such "entirely practical motivations"; nor in the admittedly unlikely event of their having financed reliably successful paranormal techniques for information transmission and trigger pulling, permitting

---

[25]S. Krippner and J. Hickman, "West Meets East—A Parapsychological Détente," *Psychic* (May/June 1974): 51–55. Later, in *Song of the Siren*, Krippner explained that the Institute existed only on paper.

recipients of government funds to provide evidence for the usefulness of such applications "in the general stream of the parapsychological literature."

Vasiliev is mentioned with approval by the four academicians, but only his popular pamphlet, *Mysterious Phenomena of the Human Psyche,* is cited, and there is no reference to his only published scientific monograph, *Experiments in Mental Suggestion,* here introduced as *Experiments in Distant Influence* [Ed. Note: the 1976 title of this book] and which he was anxious to have published in the West.

Vasiliev's book remains a classic in the field of parapsychology. That there should be some criticism of the work is inevitable, and as already indicated obvious queries affect the statistical computations. Since it has proved impossible to elucidate the origin of the arithmetical mistakes, the only proper course seemed to me to reproduce Vasiliev's tables exactly as they were published in the first edition, to draw the reader's attention to these discrepancies, and leave him to form his own conclusions. The only liberty I have taken is to make a single change in the text. On p. 106 [now page 65] (English first edition, p. 73; Russian text, p. 95 and p. 96) two different results, viz. 0.53 and 0.72 respectively, are given for the same t-test. 0.72 is correct for the figures supplied, and I have made the appropriate alteration.

The following is the list of queries which I sent to Professor Terentieva, to Eduard Naumov and Larissa Vilenskaya, who kindly forwarded them to Dr. P. Guliaev.

Table No.	Russian Version Page No.	English First Edition	American Edition	Comments
5	105	81	116–117	Should mean be 3.14?
6	107	83	121–122	Should mean first table be 5.11 instead of 4.05?
7	108	84	123	Would then be different.
8	109	86	124	Last table, S.E.M. 2, not 1.99? Small difference but seems clear.
9	111	87	125	First table—second mean, not 5.38 but 3.88 and $\sigma = 2.12$.
10	112	88	126	Would then be different.
13	122	95	139	Calculating from English version, should M = 6 mins. 26 sec.?

95 and 96	73	106	Same formula and figures given with different results, i.e.: 0.53 p. 95, 0.72 p. 96 (Russian version)

The errors would hardly seem to affect seriously either the claims that mental influencing occurred, or that metallic screening failed to inhibit the distant influencing effect. On the other hand, the experimental runs were relatively short. It would, in my view, be wrong to slur these inaccuracies. I acknowledge again my indebtedness to Dr. P. Briskin for checking the numerical data. I would also like to thank Mr. W. Kugel for confirming that the same errors occur in the German translation.

I am also grateful to Mr. Kugel for calculating at my request how long a human subject, in view of the necessary oxygen consumption, could remain conscious in Vasiliev's hermetically screened metal chambers. Mr. Kugel's calculations indicate that, given the slight amount of work performed by experimental subjects (squeezing a rubber ball), a subject could have remained conscious for five to seven hours in the iron chamber, and for eleven to fourteen hours in the lead chamber. These calculations indicate that such reservations regarding the experimental methodology in hermetically sealed screening chambers without breathing apparatus are probably groundless.

A possibly more telling criticism has recently been leveled against Vasiliev's experiments by Mr. Scott Hill at a conference devoted to the physical phenomena of psychical research.[26] Dr. Hill suggested that experimental subjects might be expected to go to sleep in a small, relatively airless room without benefit of sleep induction, hypnotic or telepathic, and that failure to control this variable in his view casts grave doubts on the validity of some of the experiments conducted under screening conditions.

Professor John Taylor has suggested that the efficiency of the screening properties of lead and iron chambers such as those used in Vasiliev's experiments might be questioned: the range and types of electromagnetic radiation is very large, depending on wavelength, and a room shielded against a particular wavelength might not be impervious to others.[27] Vasiliev himself was aware of the fact that electromagnetic waves of low frequencies and correspondingly great lengths are not completely absorbed by iron and lead walls of the 1–3mm thickness of the chamber walls he used, and that the transmission

---

[26]Conference on Bio-energetics and Psycho-kinesis, held at City University, London, organized by the Institute of Parascience, September 12–14, 1975. At the time of writing, the published proceedings are not available.

[27]John Taylor, *Superminds,* Macmillan, 1975, pp. 36/7

of telepathy by a low frequency electromagnetic field is not totally eliminated.[28] It remains to be seen if Professor Taylor can substantiate his hypothesis that telepathy is mediated by low frequency radiation of wavelengths between 60 and 300,000 km.[29]

Vasiliev's experiments as described in this book are among the few classical investigations of telepathy. They are open to some criticism, but they represent pioneering work of the utmost importance, and they should be repeated using more modern apparatus and more sophisticated statistical and experimental procedures. As Vasiliev says in the Latin tag introducing the book: "I did the best I could, let those who can do better."

Anita Gregory
School of Education,
Polytechnic of North London
September 15, 1975

---

[28]See below, Chapter 10, p. 133.
[29]Taylor, *op. cit.*, p. 139.

Editor's Note

The photographs and drawings in this book are presented here as documents of Vasiliev's work. They are reproduced from the best available prints, but are not intended as high-quality illustrations.

Russian letters employed in the plates and their Roman equivalents:

Russian	Roman
Б	B
В	V
Г	G
З	Z
И	I
П	P
ф	F

Accordingly, the round room Б (Figures 18 and 22) is referred to in the text as B, and the small side room В is referred to in the text as V, as explained in the legends.

Feci quod potui faciant meliora potentes
I did the best I could, let those who can do better

In this book are presented in outline the results of investigations which the author has pursued with his collaborators in the field of so-called mental or wordless suggestion. The author has studied questions connected with mental suggestion of acts of movement, visual images and sensations, sleeping and waking. Original experiments, both with and without hypnosis, are described, and there is an account of the so-called electromagnetic theory of mental suggestion and of the present status of the problem of the energetic nature of these phenomena.

The book is intended for biologists, physiologists, psychologists, doctors, physicists and specialists in other disciplines.

Preface

In this book I am submitting for the consideration of readers the experimental investigations of telepathic phenomena that I have pursued from time to time for forty years, often alone, but frequently also in collaboration with my colleagues. I first began these studies at the Leningrad Institute for Brain Research at the suggestion of the founder of the Institute, Prof. V. M. Bekhterev, and continued them on my own initiative after his death in 1927.

The author joined the Institute for Brain Research as a research physiologist in the autumn of 1921, at the very time when V. M. Bekhterev was enthusiastically collaborating with V. L. Durov, the famous animal trainer, in investigating the effects of mental instruction signals on performing dogs. V. M. Bekhterev attached great importance to these investigations, and devoted to their detailed description a long paper in the scientific *Proceedings* of this Institute.[1] The results obtained by his collaborators working on the same problems were also published in these *Proceedings*.[2] I also had an opportunity of participating in these experiments.[3]

In the same years Bekhterev published the results of his investigation of the effects of mental suggestion on an 18-year-old girl who was extremely suggestible. (Her task was to guess the selected article which was among 7–12 articles on a table.)[4] In the spring of 1922 Bekhterev organised a special Commission for the Study of Mental Suggestion, attached to the Institute for Brain Research, for the continuation and elaboration of his experimental studies. Among the members of this Commission were psychologists (A. K. Borsuk, N. D. Nikitin, V. D. Rabinovich), medical hypnotists (V. A. Finne, N. A. Panov), physiologists (L. L. Vasiliev, V. M. Karassik), physicists (A. A. Petrovsky, V. A. Podierny), a philosopher (C. I. Povarnin) and others. The members of this Commission devoted their attention primarily to the study of two types of phenomena: 1. The effect of mental suggestion on hypnotised human subjects; 2. psychophysiological effects of a magnetic field on hypnotised subjects. The results were submitted to V. M. Bekhterev, who deemed them of sufficient importance to warrant including the Reports of the members of the Commission in the programme of the Second All Russian Congress of Psychoneurology which took place in 1924, in Petrograd. On January 8th, at the morning session of the sub-committee of the Congress on hypnosis, suggestion and psychotherapy, the following reports were submitted by members of the Commission for the Study of Mental Suggestion:

1. A. K. Borsuk: *The present position of the problem of mental suggestion abroad.*

2. V. A. Podierny: *The study of inter-cerebral induction and perception* (experimental findings of the Commission).

3. L. L. Vasiliev and V. N. Finne: *Concerning the question of the psychophysiological effects of a magnetic field* (experimental findings of the Commission).

In addition there were two further relevant papers:

K. I. Platonov: *The force of visual fixation.* This address was accompanied by a demonstration of mental suggestion of sleeping and waking on a patient, Mrs. Mikhailova, and

P. V. Kapterev: *Concerning the problem of the nature of hypnotic phenomena.*

A special paragraph of the Resolution of the Congress was devoted to noting the necessity for further experiments of so-called mental suggestion, and expressing the desirability of participation on the part of Russian scientists in the International Committee for Psychical Research which had recently (1921) been founded abroad.[5] This Committee, founded at the initiative of the eminent French physiologist Charles Richet included regional (national) committees from nineteen European countries, and America; in the ensuing years the national committees continued to increase in number, and by 1935 there were twenty-six such committees. The resolution calling for collaboration with the international body was soon implemented. The General Secretary of the International Committee, Carl Vett, visited the Soviet Union with the proposal to affiliate a Russian Committee for Psychical Research[6] to the International Committee. At the initiative of A. V. Lunakharsky, then Minister of Education, such a Committee representing the U.S.S.R. was formed. At the beginning the Committee included Academician V. M. Bekhterev, Academician P. P. Lazarev, A. E. Kotz (Director of the Moscow Darwin Museum), and the psychiatrist G. V. Reitz who had an international reputation for his work on mental suggestion. After the death of V. M. Bekhterev, I, as his former student and follower, was included among the members of this Committee.

The All Russian Congress of Psychoneurology not only acknowledged the value of the work carried out by the Commission, but recommended its development. However, by that time the special Commission attached to the Institute for Brain Research had ceased to function; it was in fact dissolved at the end of 1923, that is just before the opening of the Congress. The implementing of the Resolution of the Congress was entrusted to the Society for Neurology, Reflexology, Hypnotism and Biophysics which was attached to the Institute for Brain Research. At a plenary session of the Society in March 1926, presided over by V. M. Bekhterev, the following Reports were presented:

1. A. K. Chekhovsky: *Concerning the problem of direct thought transmission over a distance* (experimental).

2. L. L. Vasiliev: *The biophysical foundations of direct thought transmission* (theoretical).

Prof. L. L. Vasiliev's paper, expounding a materialistic approach to the phenomena of mental suggestion was published, in popular form, in the journal *Science News* (No. 7, 1926).

As a result of the very first report, the Board of the Society instructed a group of its members to verify and experimentally duplicate the results obtained and the methods employed by the author. These control experiments did not yield any definite results; nor did further experiments conducted with the participation of Academicians V. M. Bekhterev and V. F. Mitkevich on Dr. K. I. Platonov's patient: the subject, Mrs. Mikhailova, had by this time developed a tendency to fall spontaneously into an auto-hypnotic state which in effect prevented a continuation of the experiments. However, a series of experiments with unambiguously positive results was conducted in the same year with the subject Mrs. Kouzmina, performed in the Department for Nervous Diseases of the Twenty-Fifth October Memorial Hospital. (These experiments are referred to again in Chapter 4.)

In October 1926 the Society appointed a special Commission, to be called the Experimental Commission on Hypnotism and Psychophysics, for a more systematic exploration of mental suggestion and allied biophysical problems. The Chairman of this Commission was V. M. Bekhterev and its members, in addition to those already mentioned who had addressed the Second All Russian Congress of Psychoneurology, were two physicists (Prof. A. A. Petrovsky, Prof. B. L. Rosing), two physiologists (Prof. G. P. Zeleny, I. F. Tomashevsky), three medical psychiatrists (Prof. K. I. Povarnin, G. V. Reitz, A. V. Dubrovsky), and some others. This committee, which ceased functioning after the death of V. M. Bekhterev in 1927, held sixteen meetings during the two years of its existence. Its members carried out the following experimental and methodological investigations:

1. Objective methods of analysis and control of "spontaneous" manifestations of mental suggestion were devised (Borichevsky).

2. Hundreds of experiments of "guessing of (pre-selected) visual images" (Reitz).

3. Experimental researches on manifestations of supposed recognition (Borichevsky).

4. A series of "guessing" experiments with the subject P. which gave negative results (Reitz and Dubrovsky, at the suggestion of V. M. Bekhterev).

5. A series of experiments on neuromuscular hyperexcitation using the hypnotic methods of Charcot, and by means of the conditioned reflex method (Vasiliev, Podierny and Finne).

6. A number of investigations recording breathing, pulse and blood pressure while punctures were made in various parts of the body, hands and legs (Vasiliev and Dubrovsky, in conjunction with demonstrations by To Rama).

7. Numerous experiments on the effects of placing a hand on a mobile metal needle (Vasiliev and Dubrovsky, at the suggestion of V. M. Bekhterev).

In the summer of 1928 I had the opportunity of going to Germany and France on a scientific mission. In Paris I familiarised myself with the experiments conducted by the International Metapsychic Institute founded by Charles Richet, and in Berlin with the work of the Institut für metapsychische Forshung which had just been founded. I also established scientific contacts with outstanding foreign researchers into telepathic phenomena—E. Osty, A. Rouhier, K. Bruck and others. My correspondence with the workers in these two centres of parapsychological studies continued for a long time and greatly helped me in my ensuing studies on mental suggestion.

In 1932 the Institute for Brain Research founded by Bekhterev and then headed by the well-known psychiatrist Prof. Ossipov (who did not recognise the possibility of the existence of telepathy) received an assignment to initiate an experimental study of telepathy[7] with the aim of determining as far as possible its physical basis: what is the wavelength of the electromagnetic radiation that produces "mental radio", the transmission of information from one brain to another, if such transmission in fact exists. It must be conceded that in those years (1923–1933) widespread credence was given to the experiments of the Italian physicist Cazzamalli who claimed to have discovered brain waves a centimetre in length. The experiments of Cazzamalli were deemed reliable proof of the electromagnetic theory of telepathic suggestion.

The Director of the Institute for Brain Research accepted this assignment, and its scientific execution was entrusted to me. Self-contained premises inside the Institute were allocated for the establishment of a laboratory equipped with the necessary physical and physiological equipment. The work was carried out with the scientific collaboration of I. F. Tomashevsky (physiologist) and A. V. Dubrovsky (medical hypnotist) and R. I. Skariatin (physicist and engineer); the physiologist G. U. Belitzky also took part in one series of experiments. Experts on radio technology eminent at that time were included as consultants on the physical side of the work; one of these (Academician V. F. Mitkevich) accepted the possibility of telepathic communication and himself carried out experiments on mental suggestion, applying simple techniques devised by himself.[8]

Prof. Shuleikin was another of the consultants; as opposed to Academician Mitkevich he was most sceptical about the entire issue, and rigorously criticised all our experiments, which proved most helpful. The working team of the laboratory never spared any efforts in attempting to satisfy the most demanding conditions of this hard-headed critic.

The experiments extended over five and a half years, up to 1938, and a large amount of experimental material was collected which was embodied in three reports by the authors:

1. The Psychophysiological basis of the phenomena (1934).

2. Concerning the physical basis of mental suggestion (1936).[9]

3. The effects of mental suggestion on acts of movement (1937).

The main findings of these experiments were concerned with the physical nature of "the factor" transmitting mental suggestion, and the results proved to be surprising even to those who were carrying out the experiments. Contrary to the predictions of the electromagnetic theory a most careful screening by means of a Faraday cage of the agent who was "sending" the suggestion, or of the "percipient" who was "receiving" it, or of both did not interfere with the transmission of mental suggestion in any of those instances where such suggestion was manifestly effective without screening. This unforeseen and at first confusing discovery cast doubt upon the electromagnetic theory of telepathic phenomena.

Inevitably, during World War II that deluged our country, and in the ensuing years, our studies of mental suggestion were interrupted. They were similarly handicapped for the time being in the Western European countries. The International Committee for Psychical Research was disbanded. The organisational centre of such studies moved from Copenhagen to New York (the Parapsychology Foundation Inc. exists since 1951) and the scientific contact of

Soviet scientists with foreign students of parapsychology was totally disrupted. The author knew nothing of what had been done abroad in the field of mental suggestion since the end of the '30s, and foreign scholars were burning with anxiety to find out whether any work of this kind was being done in Russia. In 1956 I established contact in writing with R. Warcollier, President of the Paris Institut Métapsychique, from whom I received some information as to the state of parapsychological studies abroad.

In 1959 the State Political Printing House published a big edition of a booklet of mine entitled "Mysterious phenomena of the human psyche". One of the chapters, called "Does mental radio exist?" is devoted to a brief survey of the present state of the problem of telepathy of which no mention had been made in the Soviet press for twenty years. The publication of this work achieved its aim: articles began to appear in the press about "Biological radio communication" including some comments by the engineer, B. B. Kajinsky;[10] Prof. P. I. Guliaev[11] included in one of his works a chapter entitled "The electromagnetic study of the brain", in which be gave his readers some information about telepathy.

At the end of 1959 and at the beginning of 1960 R. L. Kherumian, a member of the Paris Institut Métapsychique sent me two articles which had appeared in French popular scientific journals.[12] These two articles described in detail the sensational experiments on mental suggestion which were believed to have been carried out in the summer 1959 aboard the American submarine *Nautilus*.[13]

This experiment showed—and herein resides its principal value—that telepathic information can be transmitted without loss through a thickness of sea water, and through the sealed metal covering of a submarine—that is through substances which greatly interfere with radio communication. Such materials completely absorb short and partly absorb medium waves, the latter being considerably attenuated, whereas the factor (still unknown to us) which transmits suggestion penetrates them without difficulty. These results were obtained by the Americans a quarter of a century after our above mentioned data obtained in the 'thirties, and fully confirmed them. The only improvements of the American experiments over ours were that the telepathic influencing spanned longer distances and overcame greater physical obstacles, i.e. the great thickness of sea water together with the metallic covering of the submarine.

This totally unexpected foreign confirmation of our twenty-five years old experiments compelled me to make them known to a wide circle of scientific workers. The first report was made in April 1960 at the Leningrad Conference to commemorate the anniversary of the Discovery of Radio, held in the House of Scientists.[14] Our old experiments had by no means lost their interest: quite the reverse. With the passing of years their value had increased—they had stood up to the most acid test: the test of time. I had to repeat my report several times, in Leningrad and Moscow, and each time it created a great deal of interest.

Should one, or should one not, accept telepathic phenomena as definitely proved? One thing is obvious, they can no longer be ignored, they must be studied. Mental suggestion is now being investigated all over the world. It is necessary for people in the Soviet Union to know what has already been done, and what is being done, abroad, and above all we must conduct our own researches into these matters.

This is why, in 1960, the University of Leningrad considered that the time was ripe for setting up a special laboratory at the Physiological Institute of the Faculty of Biology for the study of telepathic phenomena under the direction of the author, and to publish this book.

The contents of this book represent a newly edited and combined compilation of the reports of experiments made in the '20s and '30s which I carried out jointly with some of my colleagues: V. A. Podierny, V. N. Finne (Chapters 4 and 5), I. F. Tomashevsky, A. V. Dubrovsky, R. I. Skariatin (Chapters 3, 5–9) and G. U. Belitzky (Chapter 4).

This work, in monograph form, is the substance of many years' experimental work on mental suggestion, conducted by studying the phenomena in three basic modes, namely sensory, motor and (principally) hypnotic manifestations. The greatest significance is to be attached to the experiments testing the electromagnetic theory of mental suggestion in which either the sender or the percipient was screened by metal.

In each chapter the description of my experiments is prefaced by a brief survey of the experiments conducted by other scholars who had already applied similar methods of study to mental suggestion. In order to bring the past up to date, that is to put the experiments conducted by me in the '20s and '30s in their present setting, I found it necessary to add an Appendix to Chapter 10. This will enable the reader to form a correct impression of the value of the work done in Russia in the '20s and '30s, of what in it was novel and original, what still retains its novelty, and what deserves repetition and elaboration with present day methods and techniques.

Chapter 1 of this book was written more than a quarter of a century ago. I have deliberately refrained from revising it in the belief that this will enable the reader to form a clear picture of the progress of the subject of mental suggestion in the last 25 years by comparing Chapters 1 and 10. A more detailed historical outline of the development of studies in telepathy is available in my popular science essay called "Mental Suggestion at a Distance" (State Political Printing House, 1962).

The substance of Chapter 3 was written by the engineer R. I. Skariatin who was a collaborator in our studies from 1934 onwards. It might now be possible to replace the wireless installation constructed at that time by a more modern one: however, this has still not been done by anyone performing similar tests.

Some unpublished material relating to the study of mental suggestion in the U.S.S.R. and extracts from little known Russian and foreign works that are of considerable interest for the history of the subject are included in the Appendix.

Critics often used to say to me: "You ought to begin by proving first of all, irrefutably and incontrovertibly, the existence of mental suggestion—only when you have done that should you start to think of studying and describing its physical nature, characteristics, and so forth." It is easy to forecast that similar remarks will be made about this book. The author is unable to agree with this point of view: such a policy, in his opinion, would not expedite but rather retard progress in this subject.

Even supposing that any of the series of experiments referred to by me, taken by itself, was insufficient to establish the existence of mental suggestion, nevertheless all the series of experiments taken together, particularly the recent quantitative experiments (see Chapter 10), make the actual occurrence of mental suggestion highly probable. This is an adequate justification for pursuing the investigation of these phenomena, just as though their existence had been finally established, without waiting for their universal acceptance.

In fact it is just in the course of conducting such experiments that one may find out all those conditions necessary and sufficient for the unhampered experimental production of mental suggestion phenomena: this would be the best proof of their actual occurrence. There are many parallels in the history of science. For example, the study of various physiological effects, such as the practical applications of hormones and vitamins, came long before these substances had actually been isolated in their pure form and synthesised—in other words before their existence had been proved beyond doubt.

It is a very pleasant duty to express my great indebtedness to Prof. Alexander D. Alexandrov for his assistance in organising the laboratory for my experiments on mental suggestion and in publishing this book. I thank Prof. P. V. Terentieva for her kind assistance in making the statistical calculations necessary for the quantitative material, and D. V. Zambritzkaya and O. L. Petrovitcheva for their collaboration in the preparation of the manuscript for publication. Finally I must not fail to express my heartfelt gratitude to R. L. Kherumian for sending me numerous foreign parapsychological publications.

L. L. Vasiliev
February, 1961.

1

The Origin and Development
of the Study of Mental Suggestion

Before proceeding to describe my own researches in the field of mental suggestion, it will be necessary to give a brief account of the historical background of the problem.

The phenomenon of "mental suggestion" or "direct thought transmission" or "telepathy" consists in the transmission from one person to another of various different kinds of impressions, thought, feelings and so on, and also in the possibility of thus inducing hypnotic trance; in all these cases the effects were obtained without the intermediary of words, at a distance, independently of perception by means of any of the sense organs.

The English physicist Barrett was the first to attempt a scientific verification of the reality of telepathic transmission. His first scientific report on this subject was made at a Conference of the British Association for the Advancement of Science in 1876 *("Some phenomena associated with abnormal states of mind")*. Later, in 1882, the Society for Psychical Research was founded in London. The first work of the members of this Society consisted in collecting and studying ostensibly authentic cases of the transmission of thought and feeling at a distance. The results obtained were published in 1886 by E. Gurney, F. Myers and F. Podmore.[1]

These authors were the first to propose the term "telepathy" which, literally, means "feeling at a distance", and to introduce the sub-division into *spontaneous* and *experimental* telepathic phenomena, the former occurring, as the name indicates, spontaneously, the latter being observed under experimental conditions. By experimental phenomena they meant what is now called "mental suggestion" or "mental influence": the category of spontaneous phenomena includes all the various cases in which a person feels and "sees" what is happening to another person at a distance beyond sensory reach, while this latter is undergoing severe neuro-psychic stress.

The following is one of many (688) examples taken from the London S.P.R.'s collection. One evening a Mr. U. went to spend a few days with his brother but, on arrival, instead of finding him in, only saw a note. "Instead of going to bed," wrote U, "I dozed off in a chair but suddenly, exactly at 1 a.m., I awoke and cried out: 'By God, he has fallen.' I had seen in my dream that my brother came out of the sitting room into the brightly lit hall, caught his foot on the step at the top of the stairs and fell, head first, onto his elbows and hands. I took little notice of this occurrence, and again dozed off for half an hour and woke up when my brother came in, saying: 'Ah, you are here, and I have almost broken my neck. When I came out of the big room I caught my foot on a step of the stairs and, head first, rolled down the stairs.'" Thousands of similar cases, supported by more or less reliable statements of witnesses, are described in the parapsychological literature.[2]

Manifestations of this spontaneous type cannot be repeated to order, and consequently an explanation in terms of chance coincidence cannot be ruled out. This was obvious even to the earliest students of telepathic phenomena, and it induced them to turn to attempts at an experimental reproduction of similar phenomena.

The physiologist Charles Richet was the first to conduct experiments designed to prove the reality of mental phenomena, applying quantitative methods of probability calculation to the results.[3] From that time onwards mental suggestion has become the subject of numerous and varied experimental researches. Scientists of varied disciplines became interested in the subject—mathematicians, physicists, psychologists, physiologists, neurologists and psychiatrists. A large literature is now devoted to the subject. The reality of mental suggestion was and is admitted by a number of outstanding scientists (S. Arrenius, C. Flammarion, W. McDougall, C. Richet, V. M. Bekhterev, P. P. Lazarev, K. E. Ziolkovsky, H. Berger and others).

A number of monographs on the subject of experimental telepathy were published abroad in the '20s and '30s of this century. Among them were the works of Tischner (1925), Wasilewsky (1921), Bruck (1925), Osty (1932), Warcollier (1926), and the book of the well-known American writer Upton Sinclair (1930).[4] These monographs contain vast amounts of empirical material on the telepathic transmission of images, objects, drawings, playing cards and so on. The authors of these monographs endeavoured to establish the optimum conditions, and psychic mechanisms involved in telepathic transmission and perception. The problem of the physiological and especially the physical (energetic) basis of the phenomena remains little developed by means of experiment until now.

Seventy or eighty years is usually a sufficient period of time for some idea either to be established beyond doubt or else to be entirely dismissed. In the case of telepathy we have neither the one nor the other. Present day students of the subject can be divided into three categories: some seriously consider telepathic phenomena to be definitely established facts, others (and this is a larger group) do not consider that telepathy has been proved experimentally but admit its theoretical possibility and suppose it quite probable; the last group denies the possibility of telepathic phenomena, deriving this view *a priori* from a variety of theoretical premises. "Official science still regards telepathy as non-existent" wrote Osty in one of his works.[5,6]

All this creates a general atmosphere of scepticism and pessimism surrounding the whole subject. The phenomenon of telepathy apparently exists, but we have failed to master it experimentally, and hence we have failed to establish the energetic nature of the phenomena. "Students have almost reached the conclusion", says Bozzano, "that telepathy itself presents a puzzle which becomes ever more incomprehensible the more one investigates it."[7] Such a reaction can be explained as the result of a number of factors.

In the first instance, only very few workers have, as yet, embarked on the problem. The student has to overcome a good deal of inner resistance before he can bring himself to regard the subject as worthy of study.

Secondly, despite numerous attempts, it has still not been found possible to determine the optimum psychophysiological conditions favouring telepathic transmission and perception; and it is precisely a knowledge of these conditions that would render telepathic results conclusive and of permanent value. In other words, experimenters still do not have at their disposal any certain knowledge of the best methods for studying mental suggestion.

Thirdly, the use of apparatus in a laboratory setting often interferes with the ability of the sender or agent to concentrate on the object to be transmitted to the percipient who, in his turn, is apt to become discouraged, thus reducing his ability to accept mental suggestions.

Finally, telepathic phenomena of quite different types were confused with each other until quite recently and, since such different manifestations are apt to enter into one and the same experiment, they are likely to interfere with and to mask one another.

According to present day conceptions it is necessary to differentiate between at least two types of phenomena:

1. Transmission of thought *(transmission de pensée)*—the active transmission of the "most highly charged" thought of the sender to the percipient who, in this case, plays the part of a passive recipient of the "telepatheme", the content of that which is transmitted to him: in other words, this type of telepathic phenomenon can be called mental suggestion *(suggestion mentale)*.[8]

2. Thought reading *(lecture de pensée)*—a phenomenon in which the active part is played, not by the sender but by the percipient. In this case the percipient telepathically perceives this or that content of the latent mind of the agent, for example some event of his life or some personal impression on which the agent is not at that time concentrating. This type of telepathic experience is often denoted by the term "telemnesia".[9]

This differentiation appears to be commonly accepted in the foreign literature. Some authors go even further and consider that the term "thought transmission" again conceals two qualitatively different phenomena.

The first—"transmission of thought at a short distance"—represents, according to Bozzano, "an active influence of the sender's mind upon the subconscious region of the percipient", and this is the usual subject of telepathic experimentation. The second, "telepathy proper" does not depend, or hardly depends at all, upon the length of the distance between the agent and the percipient, and represents the results of influence, whether conscious or unconscious, of the sub-conscious sphere of the agent upon the psyche of the percipient, and manifests itself mainly spontaneously in the form of so-called telepathic hallucinations.[10]

Thus, according to modern conceptions, phenomena of different types and of different degrees of complexity may enter into quite a simple telepathy experiment. For example, elements of "telepathy proper" or of "telemnesia" and even of so-called "telæsthesia" may be present in an experiment of mental suggestion.[11] This conveys some idea of the complexities that are encountered by the explorer in the realm of psychic phenomena in conducting even the ostensibly simplest experiments in mental suggestion; and it also gives some indication of the difficulties the investigator is likely to encounter when attempting to make a psychological assessment of the data obtained.

The complex nature of telepathic phenomena, our ignorance of the psychophysical conditions required to control the phenomena experimentally, our lack of sufficiently accurate and reliable methods—all these factors present unusual impediments in our attempts to determine the biophysical (energetic) nature of the phenomena.

Concerning this last question there exist at present two contradictory views.

In the opinion of the Italian researcher E. Bozzano, to whom reference has already been made, in the case of mental suggestion over short distances there is an underlying process of diffusion, from the brain of the agent or sender to the brain of the percipient, of some specific psychophysical vibrations, the nature of which is different in quality from the forms of energy known to physics.[12] However, in the view of that author, in the case of spontaneous telepathy we are not dealing with any energetic radiation at all, but with a certain "supernormal faculty belonging to the agent", i.e., "direct communication between two mentalities" *(communication directe entre deux mentalités)*.[13] This view must of course be regarded as idealistic.

The opposite view is taken by those who support the electromagnetic hypothesis of telepathic phenomena. So far as we are aware, this scientific materialistic hypothesis was first put forward by Houston 70 years ago.[14] The ideas of Houston are based on the phenomena of electrical resonance, discovered not so very long before by Herz in 1888. On this view, the neural-psychic process, accompanied in the active brain of the agent by a flow of bioelectric currents, causes in the surrounding environment electromagnetic waves of definite wavelength (brain radio waves), which reach the brain of the percipient and bring about in him an analogous process of a neural-psychic nature. In short, the phenomenon of the transmission of thought consists in a process of "intercerebral (brain-to-brain) electromagnetic induction".

A further development of this conception was assisted both by the great advances made in the field of radio technology, and by the accumulation of knowledge concerning bioelectric currents definitely associated with the process of the excitation of brain cells and nerve conduction.

V. M. Bekhterev was the pioneer of this subject in Russia. He was in no doubt concerning the reality of telepathic phenomena, and he was the first Russian scientist to invoke the electromagnetic hypothesis to explain them. It is this theory of the emanation of electromagnetic radiation by the brain that served as a working hypothesis for our investigations, and which engaged our thorough and critical attention.

2

The Electromagnetic Theory of Telepathy and Its Experimental Basis

A study of the electrical processes which take place in the cerebral cortex was begun by Keton and, in this country, by B. J. Danilevsky, in the middle of the seventies of the last century. At present the scientific literature on bioelectric currents is enormous. Here we will mention only a few of the results obtained in the course of electrophysiological studies of the cerebral cortex which have a direct bearing on the subject matter of this chapter.[1]

The foundation of these studies, and of the form in which they are nowadays carried out, were laid down in 1925 by the work of V. V. Pravditch-Neminsky. In analysing the recordings made of the rhythms of the exposed cortex of a dog he was the first to establish the several types of alternation of electric potential from the visual and the motor cortex. In 1929 Hans Berger demonstrated the possibility of recording the electrical activity of the cortex of a human being from the skin surface of the head through an intact cranium, and he recorded the rhythm of normal persons as well as those of patients suffering from various nervous and mental disturbances. Such a recording was called by him an Electro Encephalogram or EEG. The EEG of an exposed cortex differs somewhat from that measured from the scalp of an intact one because, in passing through the cranium, nerves, muscles and scalp the cortical rhythms naturally change, and become on an average ten times weaker.

The amplitude of the electrical activity of the cortex recorded in an EEG measured from the scalp is generally about 10 to 15 microvolts. There are spontaneous or automatic fluctuations of electrical brain potential which are not necessarily associated with afferent impulses from the organs of sense. The EEG of a human subject is composed of the following rhythms:

1. The α (alpha) - rhythm, frequency F = 7.5 to 13 cycles per second, amplitude A = 10 to 50 microvolts, A max. = 100 microvolts. The amplitude of the alpha-rhythm is affected by age, hunger and emotional excitation. The frequency of the alpha-rhythm is constant over a long period of life, from the ages of 20 to 70. A suppression of the alpha-rhythm can be obtained by an increased input from the sense organs. The suppression of the alpha-rhythm is most characteristically demonstrated when opening the eyes— that is, an increase of light falling on the retina.

2. The β (beta) - rhythm, frequency F = 15 to 25 cycles per second, amplitude A = 25 to 30 microvolts. The beta-rhythm is most easily elicited when the alpha-rhythm is suppressed by increased sensory input. The maximum amplitude of the beta-rhythm becomes apparent when it is recorded directly in the pre-central convolutions of the cortex.

3. The θ (theta) - rhythm, frequency F = 4 to 7 cycles per second, amplitude A = 10 to 30 microvolts. The theta-rhythm is characteristically associated with the emotions. It appears in the EEG of children when they are in a temper; in adults it is associated with aggression which also increases the amplitude.

4. The δ (delta) - rhythm, frequency F = 0.5 to .3 cycles per second, amplitude A may amount to 1,000 microvolts. The incidence of the delta-rhythm in the waking cortex is a pathological sign; its appearance in sleep is normal.

5. Rapid spikes, frequency F = 500 to 1,000 cycles per second, amplitude A = 150 to 200 microvolts. These vibrations are primarily located in the anterior temporal lobe. Some authors consider them to be due to the local stimulation of nerve cells.

What is the greatest amplitude to which the highest voltage potentials can rise, and on what does this amplitude depend?

High voltage discharge in the cortex can be explained by the following factors: 1. the total number of neurons excited at the time; 2. the synchronisation of phase (simultaneity) in the discharge of the population of neurons that create the sum-total of the discharge; 3. the rate of depolarisation of each cell (so long as the cellular potentials do not follow the *all-or-nothing* law). If it is contended that this latter factor is constant, and that each cell always emits its maximum potential, then the height of potential will be proportional to the number of cells synchronously excited. In this case greater potentials are associated with higher degrees of synchronisation: the same numerical population of cells, if discharging in a disorderly asynchronous manner, will produce a lower amplitude than if excited in phase. In high states of synchronisation it is possible to record potentials 10 to 20 times their normal amplitude.

The electrical activity of the cerebral cortex is unquestionably in some way connected with mental activity. In all cases where a person, for one reason or another, loses consciousness, there appears in the EEG a slow delta-rhythm of high amplitude. If the loss of consciousness terminates in death the oscillations generally become weaker, eventually decreasing to zero. The electro-encephalogram of any given person has certain individual characteristics, apparently connected with the distinguishing peculiarities of his psychoneural make-up. It is however impossible, and it probably never will become possible, to decipher from an EEG the state of mind and the sensory experience of an experimental subject, i.e. what exactly he is thinking, feeling and experiencing at that time.

What is important for our present subject is that rhythmic electric currents emanate from the brain and can be detected at the surface of the scalp; it follows of necessity that the human head must be surrounded by an electromagnetic flux. The question that arises is whether present-day radio-techniques are sufficiently sensitive to record the brain's electromagnetic waves.

The whole subject was substantially advanced by Academician P. P. Lazarev (1920) who wrote: "We must, thus, consider the possibility of catching in space a thought in the shape of an electromagnetic wave; this would seem to be one of the most interesting problems in the whole of biological physics. One must, of course, realise the immense difficulties that, in principle, stand in the way of detecting such waves. Many years' strenuous work will be required before isolating and demonstrating these phenomena, but they are inescapably forecast by the ionic theory of stimulation. The transmission of thought processes through space provides a firm basis for an explanation of the phenomena of hypnosis, and this concept is of the greatest interest both from a theoretical and from a practical point of view."[2]

Having taken into account the cortical rhythms (10 to 50 cycles per second) and the speed of diffusion of electromagnetic waves (300,000 km per second), Lazarev inferred the wavelength of brainwaves to be 6,000 to 30,000 km. It has, so far, proved impossible to record experimentally such long waves around a human head.

Academician V. M. Bekhterev wrote on the question of the nature of the energy involved in mental suggestion: "This question certainly requires detailed investigation but, in view of the fact that a neural current is accompanied either by electro-negative oscillation or by active current, and that such an active current itself produces fluctuation on account of the effects of the accompanying ionic phase of decomposition on the phase of restoration of the nervous substance that ensues, there are grounds for believing that we are here dealing with a manifestation of electromagnetic energy, and most probably with Herzian waves."[3]

The above quotations show that, as opposed to Academician Lazarev, V. M. Bekhterev considered that mental suggestion is effected, not by kilometre low-frequency fields, but by short high-frequency electromagnetic waves.

This assumption of Bekhterev's found support in the experiments of the Italian neurologist Cazzamalli[4] who, in his experiments conducted from 1923 to

1925, used a screening chamber built according to the principles of a Faraday cage. In that author's opinion, the covering of sheets of lead (1.5 mm in thickness) served completely to protect the inner surface of the chamber from exterior electromagnetic fields that would otherwise penetrate it. The experimental subject who was susceptible to being put into deep hypnotic trance by means of verbal suggestion was placed in the chamber, together with a radio-valve receiver, with a loop or lineal antenna placed at a distance of 50 to 70 cm from the head of the subject.

The radio receiver, tuned to a short wavelength (from 10 to 100 m, and in some experiments from 0.7 to 10 m) was connected by a wire to a pair of ear phones placed on the head of the experimenter, who was either outside the chamber or, in some cases, inside it. Before the experiment all the equipment was tested. The subject was then put into the chamber and the apparatus was checked again, while the subject was reclining. So long as the subject was awake, the radio valve receiver emitted no signals but, as soon as he fell into hypnotic trance and, under the influence of suggestion, began to hallucinate, the telephone began to record divers sounds, thus indicating the formation of radio waves. These sounds resembled either the usual radio-telegraphic signals, or intermittent whistles, or the sound of a violin and, according to the experimenter's assertion, these sounds differed from the type of noise made by some defect in the installation. The sounds increased with an intensification of the suggested hallucinations, and died away as they waned. The instant the subject awoke the sounds ceased and, as soon as the hypnotic condition was induced again, the noise started once more.

Such experiments, carried out with the assistance of experienced radio technicians on a number of subjects, generally patients suffering from some form of psychoneurosis, convinced Cazzamalli that the human brain, when under the influence of highly emotionally tinged psychoneural states produced under hypnosis, radiates into the surrounding space electro-magnetic energy in the form of aperiodic and detectable radio waves of 0.7 to 100 m.

A number of weighty objections were levelled against these first experiments of Cazzamalli's from a methodological point of view. Two authoritative physicists, Breno[5] and A. A. Petrovsky[6] considered that the inferences drawn by the Italian scientist were not entirely conclusive for the following reasons: "As a consequence of the high sensitivity at which it is necessary to set the radio-receiver the slightest shift of parts of the apparatus, the approach of a hand, or even the changes occurring inside the battery, can suffice to cause the generation of radiowaves. This often takes place in radio practice." When one uses radio sets of the type employed in his work by Cazzamalli, "the experimenter is inevitably faced with a generator of electromagnetic waves. These waves, continuously reflected from the metallic walls of the chamber and interfering with one another may greatly complicate the picture of the phenomena." Besides, "Since the subject is within the field of electromagnetic waves emitted by the radio receiver itself, and he himself is a conductor, electrical currents of a vibratory nature must necessarily be elicited as a result of which his body will create its own electromagnetic field which in its turn will react on the receiver." For the human body the conditions of resonance, when this effect is apparently most pronounced, occur with a wavelength of about 3 m.[7] When the waves are longer or too short the system does not resonate and the manifestations of modulation, by which Cazzamalli's results can be explained, are weakened.[8]

In the experiments conducted from 1926 to 1933 that followed, Cazzamalli considerably improved and perfected his methods, taking into account a number of the criticisms raised. The human experimenter, listening to waves through a telephone, was replaced by galvanometric recording of waves.[9] In his latter work[10] Cazzamalli used an electromagnetic oscillograph equipped with intensifier and apparatus for automatic photographic registration of the effects onto cinéfilm.

In the first series of experiments the subject, who was lying on a couch in the screened chamber, was not permitted to move. When he imagined with vivid intensity persons or events of a deeply emotional significance for him, or when he fell asleep and had vivid dreams, or, finally, when he was subjected to suggested hallucinations, the oscillograph often began to register the electromagnetic waves picked up by the receiver.

In another series of experiments the subject was instructed to relax as much as possible and to induce in himself a state of "mental passivity." This state would suddenly be interrupted by the experimenter who would suggest to the subject that he should think about people or events which evoked intense emotions.

After a short latent period the indicator of the oscillograph began to move, the oscillogram would be recorded in the form of a number of rhythmic damped oscillations of definite frequency; in other instances the vibrations were a-periodic, i.e. of different wavelengths. Figure 1 represents an oscillogram recorded during an experiment with a subject who was instructed to imagine the picture of a battle during the first World War in which he took part,[11] (from the work of Cazzamalli, 1933).

The eliciting of vibrations in response to a stimulus was called by the author a "psycho-radiational reflex." It is similar to the scientifically well known and long established "psychogalvanic reflex" with this difference, however, that here, in response to the afferent stimulus, it is not a bio-electric current but a number of electromagnetic rhythms that are registered by the apparatus at a certain distance from the subject.

The radio brainwaves detected in the experiments here described are considered by Cazzamalli to be the energetic factors underlying telepathic transmission of thought. The radio-waves emitted by the brain of the agent or sender penetrate into the brain of the percipient and stimulate such of his cortical centres as are tuned to the corresponding wavelength ("brain radio," "intercerebral induction").

The work of Cazzamalli was not entirely novel. As always, there were fore-runners. In the first half of the last century the famous British physicists Davy and Faraday demonstrated that at the discharge of an electric (Torpedo) fish the magnetic poles deviated and the steel needles became magnetised according to the same laws to which ordinary electric currents are subject. This meant that "animal electricity" as it was then called, can act at a distance.

Among attempts to detect an electric field created by the human organism mention should be made of the work of Hedweiler[12] who, in his experiments, used a quadrant electrometer connected to a receiving plate. Movements of a human being near the plate caused a change in the electrometer readings. Clearer results of the same type were later established by the German physicists Sauerbruch and Schumann[13] who found that the contraction or even the mere tension of the muscles of the upper and lower extremities of a human subject were accompanied by the generation of an electric low frequency field which could be detected at a distance of 3 m. These experiments will be referred to again in Chapter 4.

Such low-frequency fields can be generated by the ordinary vibration of the active skeletal muscles and, as is well known, the rhythm of muscular vibration does not exceed 40 to 50 cycles per second. But what bioelectric processes could generate high-frequency cerebral waves? To this day electrophysiologists have not encountered such high-frequency electric vibrations either in the EEG or in the ECG which could independently generate the Cazzamalli waves of 10 cm to 1 m wavelength.

Some authors—B. B. Kajinsky, A. V. Leontovich and G. Lakhovsky—tried to overcome the difficulty by proposing the ingenious "histophysical" hypothesis.[14] Kajinsky started with the following premises. In his time Apati formulated the idea that the conductors of the electrical motor currents are the unsheathed nerve fibres, whereas the myelinated nerves act as insulators. This was experimentally confirmed by Bethe. Bekhterev suggested that the ends of dendrites play the part of condensers in the process of synaptic transmission. All this was advanced by Kajinsky as an argument in favour of his theoretical conception of the vibratory nature of the neuron system.

For the transmitter system it is sufficient to have inductive capacity. Kajinsky believes that the afferent excitation is generated by the membranes of the nerves, whereas the efferent excitation represents the neuron circuits. Kajinsky draws his conception of a receptor from histology. In his view the "Krause corpuscles" play the part of antennæ, whereas the part of detectors and intensifiers is played by the ganglionic cells of nerves (which, as he says, work like triodes). Since the basic frequency of the nervous centres is (by Verworn's measurements) 12 cycles per second, the corresponding wavelength is equal to 25,000 km. We know from radio that an open symmetrical vibrator emanates a wave four times longer than the vibratory length itself. It follows that the vibrator in the brain must not be less than 6,000 km in length which is a difficult assumption.[15] Kajinsky believes that in the nervous system closed circuits are likewise present.[16] The basic deductions of Kajinsky resolve themselves into the following:

1. Both open and closed vibratory currents of the neurons may be interconnected by mutual induction, and the effect in such interaction is marked by the occurrence of combined oscillations as a result of which a detectable electromagnetic wave is generated.

2. The entire nervous system represents one combined system of neuron chains that generate one composite electromagnetic wave.

3. Every thought is accompanied, in the central nervous system, by the generation of electromagnetic waves.

4. The electromagnetic waves of one brain can be afferently received by another brain and excite therein a corresponding psychoneural experience.

In our view the histological premises advanced by Kajinsky are not convincing, the whole of the reasoning being based on the analogy between the morphological structure of the nervous system and the structure of physical receivers and transmitters. Such an analogy may be purely formal, and quite generally analogical reasoning is unreliable.

However, the theories advanced by Kajinsky aroused the interest of the eminent physiologist and hypnotist A. V. Leontovich. Starting in the year 1928 Academician Leontovich tried to discover in the histological structure of the nervous system necessary and sufficient conditions for the generation of ultra-short radio-waves corresponding in length to those of Cazzamalli. The different structural components of nervous tissue (pericellular fibres, neurofibrils, cell extensions with their lamina terminalis), were viewed by Leontovich as conductors of bioelectrical current possessing self-induction and capacity. From this point of view the nervous system is considered as a collection of a tremendous number of radio installations or sets of microscopic size, capable of generating and receiving ultra-short radio-waves. Thus radio-waves generated by pericellular fibres may, according to Leontovich, be intercepted by intracellular neurofibrils. By this process with the help of electromagnetic energy stimulation is transmitted inside a brain (intra-cerebral induction).[17] Micro-measurements and mathematical calculations convinced Leontovich that the length of the radio-waves generated in this case corresponds to the length of Cazzamalli's waves.

Cazzamalli's experiments and Leontovich's calculations would appear to afford reliable corroboration for the electromagnetic theory of mental suggestion. Nevertheless, in spite of its apparent simplicity, this attractive theory has to contend with a number of serious objections.

Let us assume that cerebral radio-waves in fact exist, that the brain of the sender generates them in the course of a telepathic experiment, and that such waves penetrate into the brain of a percipient. The question now arises whether their potential is sufficient to enable them to stimulate the nerve cells of the percipient? This, undoubtedly, is absolutely essential: otherwise, their action will be below the threshold of stimulation and will remain without psychophysiological effect.

This important point was first raised by Prof. V. A. Arkadiev.[18] Making use of some equations of theoretical physics he tried to work out the strength of the current and the size of the field which could correspond to such brain radiations. The results of his calculations provided little comfort: the strength of the current would be of the order of 10^{-18} ampères, whereas the energy carried over by the electromagnetic radiation of the brain is of the order of 6.54×10^{-24} erg, i.e. considerably smaller than the minimum energy of light perceptible to the most sensitive receptor—the eye (2×10^{-10} erg).

It is hardly possible that brain cells are more sensitive to radio waves than are the receptors to the stimuli adequate to excite them—the eye to light, the ear to sound, etc. Arkadiev concludes his account by stating that the size of the field or the strength of the current resulting from the electromagnetic emanations of the sender's brain are too insignificant to be capable of having any physiological effects on a percipient's brain. The author observes "The electromagnetic fields generated by the electrical installations among which modern man lives and works ought to have a considerably greater effect on his nervous system." To-day, a quarter of a century later, the same observation still holds: electric or magnetic fields do not seem to have any noticeable effects, useful or otherwise, on the human organism.[19]

To all the foregoing it must be added that the wireless waves described by Cazzamalli could hardly have the stimulating effects on a percipient's brain claimed for them, not only because they are not powerful enough, but also because of their very high frequency. According to the law of stimulation for alternating currents (the so-called square root law of Nernst), the higher the frequency of the currents and the electromagnetic fields generated by them, the less the stimulatory effect: $r = a/\sqrt{n}$, where r stands for intensity of stimulation, a for amplitude and n for the frequency of the alternating currents. The unexpected fact that high-frequency currents have no physiological effects on organisms, discovered by the Yugoslav physicist N. Tesla, is a direct consequence of this, and so is the observation made in a number of laboratories concerning the physiological effect of a high-frequency electromagnetic field on the functional condition of the brain, the nervous system and the whole organism. High frequency (metre) radio-waves of sufficient power are apt to kill the neuromuscular apparatus and indeed the whole organism without simultaneously eliciting the slightest sign of nervous stimulation or excitation. The death of the organism is due to overheating which arises as a consequence of Foucault currents. Low-frequency fields (of wavelengths of several thousand kilometres), on the contrary, produce a highly excitatory effect on isolated nerves and cause an increase of stimulation of nerves and nerve centres when *in vivo* in their natural setting in the organism.[20]

It is of the utmost importance to note that low-frequency electromagnetic fields can apparently affect the human cerebral cortex, and hence higher nervous activity. In this connection particular attention must be drawn to the work of F. N. Petrov, carried out at the I. P. Pavlov Institute of Physiology, which is attached to the Academy of Sciences of the U.S.S.R.[21]

Petrov succeeded in establishing in two healthy human subjects a conditioned defensive motor reflex to the sub-threshold effect of an electromagnetic field of low frequency (200 cycles per second). Stimulation of the fingers of the subject by means of an electrical current served as stimulus eliciting the unconditioned response. The charge on the vibrator, placed over the head of the experimental subject was 30,000 volts. In order to establish a conditioned reflex not discernible by the subject a positive electric surface discharge was added to the electromagnetic field. The breaks between tests were 0.5 to 3 minutes.

The first conditioned response to the field from subject G (40 years old) was obtained on the 150th test. It is important to note that in this subject conditioned reflexes to felt, that is subjectively experienced above-threshold signals such as light and the sound of a bell, were only elicited with equal difficulty. The first conditioned reflex to the field of another experimental subject (25 years old) appeared on the 67th test. The conditioned reflex of both subjects became stronger after first appearing, but had no lasting stability.

Subsequently an attempt was made to produce a conditioned inhibition in response to an electromagnetic field. To begin with, a conditioned defensive reflex to red light was elicited and established in subject G. Then the red light was, from time to time, reinforced by an electrical skin stimulation and the combination *electromagnetic field plus red light* was not reinforced. In fact conditioned inhibition was gradually established (only at the 36th test) but this inhibition again was not stable.

F. P. Petrovsky explains the extreme tardiness and instability of positive and negative conditioned reflexes in this case as being due to the fact that an electromagnetic field of low frequency is not an adequate stimulus for the receptors present in the human organism. He refers to the findings of Academician K. M. Bikov to the effect that if the stimulus, even though felt, is inadequate in intensity, the conditioned reflex either is not formed at all, or else appears sporadically and seems weak and unstable.

Recently it was shown in the Department of Physiology of Higher Nervous Activity at the University of Moscow that in a species of fish defensive and food (alimentary) conditioned reflexes can be established to the subthreshold effects of a magnetic field. The threshold of perception of a magnetic field proved to be 10 to 30 œrsted. It proved impossible to establish any conditioned reflexes to a magnetic field in most species of birds, particularly pigeons. In other species of birds the magnetic field, as in fishes, caused a sharp increase (about 4 to 6 times) in motor activity.[22]

U. A. Holodov observes: "Animals presumably do not possess special receptors for perceiving magnetic fields, but the diencephalon appears to be most sensitive to them as well as to other penetrating radiations (X-rays etc.)." The present author can support this view from personal observation. After injury to the diencephalon (other parts of the central nervous system remaining intact) conditioned reflexes established in fish to a magnetic field were disrupted; it was also found possible to establish in frogs Sechenov's inhibitions of spine-brain reflexes by using the direct reflex effect of a magnetic field on the mid-brain. U. A. Holodov draws the following important conclusion: experiments with the prolonged effects of a constant magnetic field showed that animals respond, not only to changes in its intensity, but also to constant intensity.

Experimental data (not yet published) are available showing that a conditioned reflex can be established to the sub-threshold effects of electromagnetic waves of ultra-high frequency. Possibly this physical stimulus can, like a magnetic field, act on the brain independently, that is without special receptors. It is now experimentally established that a temporary connection between cortical neurons can be made without the intermediary of a receptor, by means of repeated independent simultaneous stimulation by electric currents of sensory and motor sections of the cerebral cortex.[23] This does not contradict the reflex theory and teachings of I. P. Pavlov concerning conditioned reflexes. The facts do support the conception of "brain radio."

This conception is further supported by the experiments of the Canadian scientist W. Penfield.[24] He showed that, by electrical stimulation of specific portions of the cerebrum, it was possible not only to establish conditioned reflexes but also to elicit different types of subjective experience. Penfield carried out his experiments on epileptic patients. A generator of electrical impulses of 60 to 300 cycles per second frequency was employed as stimulating apparatus, stimulation being carried out by microelectrodes placed near the pathological focus but in a locus not affected by the disease, in the temporal lobe of the cortex. During stimulation the patient would tell that he, so to speak, re-lived some past episode of his life. If the electrodes were shifted by as little as 1 mm this episode was superseded by another. The patient was able to describe his subjective experiences to the experimenter while fully aware of his actual surroundings and the fact that he was lying on the operating table.

Thus mental experiences which were in the past created in the subject's mind by the usual reflex processes could be caused to re-appear without receptors or reflex arcs—by direct electrical stimulation of quite specific neurons of the cerebral cortex. It is natural to suppose, as a next step, that in "direct thought transference" the same neurons are stimulated, not by local currents but by electromagnetic waves of specific lengths emitted by another brain.

Supporters of the electromagnetic hypothesis of mental suggestion attach equal importance to experiments in which the suggesting "sender" and the "percipient" were placed in a metallic chamber. One would expect *ex hypothesi* that the metal screening would appreciably reduce the effect of mental suggestion if such effects were in fact transmitted by electromagnetic waves. So far as we know, the first experiments of this type were in fact conducted by B. B. Kajinsky and V. L. Durov on trained dogs, in the mid-twenties.[25] In 61 per cent of the experiments in which the sender was screened, the mental effect on the animals ceased, but in 39 per cent of similar types of experiments the effects continued to manifest themselves—a somewhat inconclusive result.

Another series of experiments was conducted by the hypnologist T. V. Gurstein, together with B. B. Kajinsky in 1926. A chamber with double walls, with a floor and ceiling made out of sheet brass and iron was used. The chamber was placed on insulators and could be earthed through a commutator (two-way switch). In the opinion of these investigators, this "circuiting," i.e. the earthing of the walls of the chamber, increased its insulating properties.

The percipient was placed in the chamber and Gurstein, while remaining outside the chamber but in the same room, would look at some object in his hands and ask the percipient to describe it; or he would ask the percipient whether he smelt anything (whilst himself sniffing some eau de cologne), or whether he felt anything (whilst pricking his own hand with a needle). 14 such experiments were carried out from which, however, it is hard to draw any conclusions as to the importance of screening: the percipient responded accurately both when the chamber was earthed and when it was not. For example, Gurstein, holding in his hand a red cross, asked "What do you see?" The answer—the chamber being earthed—was "A cross." In the course of the same experiment, in response to the question "What colour do you see?" the reply was "dark red"—the chamber not being earthed.

Another series of experiments involving screening was carried out by T. V. Gurstein together with L. A. Vodolazov in 1936, that is when we were nearing the completion of our investigations at the Leningrad Institute for Brain Research. In 1937 Dr. Gurstein reported the results of his experiments to a meeting of the Moscow Society of Psychiatrists and Neurologists.[26] This series of experiments was more elaborate than the previous one and deserves closer attention. It will be referred to again in Chapter 4 which is devoted to the effects of mental suggestion on motor activity.

In that chapter we will also discuss another attempt to demonstrate experimentally that screening by means of metal delayed the transmission of mental suggestion. This work was carried out by one of the collaborators of Academician Lazarev, S. J. Turligin. He described his results at the 1939 meeting of the Biophysical Section of the Moscow Society by naturalists, in a paper entitled "On the radiation of the nervous system of man." For our present purpose it will be sufficient to note one of the conclusions of the paper, namely that wordless suggestion is affected by electromagnetic waves, one of the wavelengths of which, measured by means of a directional lattice, constructed from parallel wires, proved to be about 2 mm or, more accurately, 1.88 to 2.1 mm.[27]

The electromagnetic hypothesis thus became amplified by a further elaboration: electromagnetic waves about 2 mm in length were supposed to play a central part in the process.[28]

Everything contained in this chapter shows that the electromagnetic hypothesis of mental suggestion, while certainly corroborated by a number of facts, nonetheless encounters a number of difficulties and contraindications which have not yet been eliminated. In spite of this, when beginning our investigation, we were guided by this hypothesis and did not have, at that time, any doubts as to its truth.

3

The Physical Apparatus—Exploratory and Verification Experiments—Research Design

As we have seen, the central research problem consists in ascertaining as far as possible the physical nature of the energy which causes the phenomena of mental suggestion. As has been stated we took as our working hypothesis the electromagnetic theory, and consequently it seemed to us that it would be interesting to conduct a series of experiments designed to verify this hypothesis so as to establish brain radiation as a fact.

In the first instance we decided to repeat Cazzamalli's experiments, the results of which, so far as we knew, had still not been adequately confirmed by other workers.[1] We therefore decided to construct appropriate apparatus, in accordance with Cazzamalli's specifications, and we will now proceed to describe this equipment.

As has already been mentioned, Cazzamalli, in order to exclude extraneous influences (i.e. induced or extraneous electromagnetic fields) conducted his experiments in a chamber, the walls of which were covered with 1.5 mm thick lead sheeting. We decided to use sheet iron as a screening material in view of the fact that, with electromagnetic waves of the anticipated lengths (1–10 m) the nature of the material of the screen is not of great importance. Our chamber (to which we shall refer as the Faraday chamber, or Chamber No. 1) consisted of a wooden framework covered with sheets of roofing iron about 1 mm in thickness. The seams between sheets were soldered with lead. The measurements were as follows: length 2.2 m, width 1.4 m, height 2.0 m (see Fig. 2). Inside the chamber there was a bed for the use of the percipient, a table for the various pieces of apparatus and a chair for the experimenter. The chamber could be entered through a door covered with sheet iron which fitted tightly without leaving any crevices.

According to Cazzamalli's description he used receivers based on a single phase system. The receiver had a tuning range starting with wavelengths of 70 cm to 5 m. The receiver was a four valve one. The first valve served as regenerative rectifier, the second and third as low frequency amplifiers with transformer coupling, and the fourth as second rectifier rectifying the audible frequency of the current and feeding it into the oscillograph circuit.

Our aim being to verify the Cazzamalli experiments, however, we decided to deviate somewhat from a precise duplication of his apparatus for the following reasons:

First, since we did not possess an oscillograph and were in any case not certain whether we should obtain any positive results, we decided not in the first instance to employ automatic registration equipment, but instead to rely on subjective indicea—hearing—for the purpose of registering the effects, if any.

Secondly, we decided to replace the single phase system by a two-phase system with inductive coupling of the grids and anodes. This divergence was due to our inability to make the valves we had work as a regenerative rectifier on very short waves. For the same reason, the minimum wavelength of our receiving set was 2.78 m. We were unable to obtain stable results on shorter waves on any system we tried out. Single phase systems gave no stable results even on longer waves. These were the factors that led us to use the push-pull system with inductive reverse coupling.

The general layout of our receiving apparatus is shown in Fig. 3. As can be seen, the receiver had on the transformers, in addition to two receiving valves, also two stages of low-frequency amplification. Valves of the U.B.110 type were used. The oscillating circuit was inductively coupled to an aperiodic aerial stretched along the chamber and over the bed at a height of 70 cm. The reception was by head phone.

In order to test the working of the receiver and the screening properties of our chamber we built a push-pull generator working on valves of the UT 1 or similar types. The generator could be fed either directly from the AC mains supply or from accumulators. The circuit system of the generator is represented in Fig. 4. The generator had no radiation system. The measurement of wavelength at the receiver as is usual in shortwave technique was made by means of the Lecher system.

Experiments showed that the output of the generator could easily be received at any place in the laboratory, neither the receiver nor the transmitter having any aerial.

Before embarking on the Cazzamalli experiments we tested the screening properties of the Faraday chamber. The experimenter was placed in the chamber with a receiver which was tuned on the wavelength of the generator, the door remaining open. The door of the chamber was then gradually closed, the observer noting the strength of reception all the time. The test showed that when the door was tight shut it was impossible to detect any sign of reception whereas if the door was not tightly closed—if there remained even a small chink (0.5 to 1 mm)—there was instantaneous and loud reception. In that series of experiments the generator was in the same room as the chamber and at a distance of 1 or 2 metres from the latter. These checks carried out over a range of wavelengths from 2.8 to 15 m convinced us that when the door was tight shut no external fields penetrated into the chamber.

Cazzamalli had at first carried out his experiments on hypnotised subjects to whom highly emotional experiences were suggested. He subsequently detected electromagnetic fields in the case of subjects who were not in a hypnotised condition but were re-living in imagination some very stirring episode in their past.

We conducted a series of similar experiments. In some of them subject, hypnotist and experimenter were all in the Faraday chamber; the subject was on the bed and was hypnotised by some technique or another. Emotions of a pleasant and of an unpleasant nature were then suggested verbally, in response to which the subject would sometimes display violent mimicry accompanied by somatic and verbal reactions. All the time that the experiment was in progress the experimenter (R. I. Skariatin) listened in on a telephone slowly going through the whole range of the receiver.

In other experiments two persons were present in the Faraday chamber, the subject and the experimenter. The former took up his position on the bed under the aerial, and tried to re-live as vividly as possible some very highly emotionally charged events from his past. Meanwhile the experimenter was carrying out the same observations as before.

All these experiments gave negative results: not in a single case did we manage to detect any unusual sounds or noises in the telephones of the receiver. The only noticeable sound was a faint rustling, normal for any regenerating receiver, which is caused by incomplete stability of the electric processes in the receiving devices.

To what, then, should we attribute the failure of our experiments to yield any positive results? Our divergence from Cazzamalli's apparatus would not seem to account for the difference in results. The substitution of iron for lead as a screening material cannot affect the radiation (if any) of electromagnetic waves from the subject's brain, since it is known from radio technological practice that the nature of the screening material at most affects the wavelength of the radiated vibrations and the decrement of the tuned circuit. Nor could the use of a double phase system, instead of a single phase system, affect the sensitivity of the receiver since it is known that both systems possess equal sensitivity.

Waves somewhat longer than the ones with which Cazzamalli worked could hardly affect the result, since Cazzamalli obtained reception not only waves of about 1 m in length (as described by him in his second, 1929, work) but also on longer ones of a length up to 10 m (cf. his first work in 1925).

It ought to be mentioned at this point that the photographs of the receiving set-up reproduced by Cazzamalli in his 1929 book made it seem doubtful whether he carried out experiments on wavelengths of 1 m: the size of the coils giving self-induction of the oscillatory circuit of the receiver would have been quite incapable of ensuring the amount of self-induction necessary for tuning in on a 1 m wavelength, even taking into consideration only the inter-electrode capacity of the first valve. Besides, we do not know what a receiver working on a principle of reversed coupling on such a short wave would be like.

If the Cazzamalli experiments are sound, the reason for the negative results yielded by our experiments must be sought in the unsuitability of the subjects selected by us.

The fact that we obtained negative results in our attempts to repeat Cazzamalli's experiments induced us to begin to search for different physical indicea of radiation emitted by the organism. Thus, we tested experimentally the claims of one electrical engineer to the effect that he had managed to register electromagnetic radiation, emitted by a brain, with a wave 1 mm in length.[2] In order to prove his claim this engineer constructed his own special apparatus in our laboratory. No description of these experiments or of the apparatus will be given, because the results were negative.

It should be made clear right away that we did not attach any great importance to these preliminary experiments. Even had they given unambiguously positive results, and if the Cazzamalli brainwaves had been established as indubitably factual, the question would still remain open whether the Cazzamalli brainwaves and not some other region of the vast electromagnetic spectrum have anything to do with the factors productive of telepathic transmission. This is the very core of the problem and the main task was to give a reliable experimental answer to just this question.

In order to do this it was necessary to devise a crucial experiment. Our consultant physicists, Academician B. F. Mitkevich and Prof. A. A. Petrovsky, fully approved our experimental design which we devised along the following lines:

1. To select a method of mental suggestion suitable for this experiment, a method that would, as far as possible, yield homogeneous data suitable for subsequent analysis by means of statistics.

2. Having selected our method, to conduct a sufficient number of tests on subjects suitable for telepathic experiments, and to establish the existence of telepathy statistically.

3. To repeat the same experiments, using the same methods and the identical subjects but under conditions where either the subject or the sender were

placed in a chamber screened from electromagnetic fields, and hence from the alleged Cazzamalli brain waves.

If it was found that screening did not reduce the percentage of "telepathic hits" then we could establish the null hypothesis, namely that electromagnetic waves, including the Cazzamalli brain radiation, which are stopped by the screening, play no part in the process of thought transmission. In that case, by changing the material of the walls of the chamber (by replacing iron by lead, etc.) we could in future solve the same problem for other electromagnetic frequencies.

These, then, were our aims. In the literature available to us we could only find one attempt that was at all similar to ours, namely the experiments conducted in 1923 by the engineer B. B. Kajinsky jointly with V. L. Durov,[3] but these experiments seemed to us to have been methodologically unsatisfactory. This will be discussed in the next chapter.

Still not wholly persuaded of the screening properties of our Faraday chamber we decided to build a new one—a perfect screening chamber (chamber No. 2) under the guidance of Academician B. F. Mitkevich.

This chamber was in the form of a rectangular parallelopiped, 1.7 m in height, 0.62 m in width, 0.90 m in length, and consisted of an upper and a lower part. Each part consisted of a wooden framework covered with 1 mm iron sheeting. We carefully soldered all the seams and joints. The upper half "the bonnet" was suspended from a steel cable passing over a pulley block screwed into the ceiling. The weight of this upper half was counterbalanced by a weight of 16 kg, by means of which we could easily raise this bonnet or lower it onto the other part of the chamber. Along the upper edge of the lower part of the chamber there was a gutter filled with mercury. The lower edge of the "bonnet" when lowered rested in mercury, and in order to obtain optimum contact this edge was made from brass and was amalgamated. This set-up ensured the absence of cracks between the upper and the lower portion, thus securing a complete screening of the space inside the chamber from external electromagnetic fields. The chamber contained a chair and a small table for the experimenter and was lighted by a small lamp of the type used in cars, and which was worked from a nearby accumulator.

The experimenter inside the chamber remained in touch with the outside world by means of a contact arrangement which closed when he pressed a particular place of the iron cover of the chamber. This contact caused the circuit of the electromagnetic recorder to close. In addition an electromagnetic buzzer was placed on the outside of the chamber. By means of this buzzer it was possible to keep the experimenter informed of the condition of the subject, e.g. whether he was in trance or awake. Thus, the chamber was hermetically sealed by means of iron, having nowhere any cracks or conducting wires.

In order to test the hermetic sealing properties of the chamber a metallic valve was soldered into the wall of the chamber, and the pressure inside the chamber could be tested by means of a connected manometer. Testing showed that the pressure inside the chamber remained constant for a period of 2 hours.

The testing of the perfect electrical insulation was conducted as follows. A shortwave generator (described above) was placed inside the chamber together with its batteries. The observer, moving around the outside of the chamber with a receiver attempted to detect any leak of electromagnetic radiation from the chamber to the outside. A variation of this testing procedure was that the observer, placed inside the chamber with his receiver, tried to intercept signals coming from the generator which was outside the chamber. Fig. 5 shows the outer aspect of this complete screening chamber. The upper half of the chamber is raised, and the generator in the chamber can be seen. On the left, on the table, there stands the receiver, and on the front wall of the lower half of the chamber may be seen the signalling contact apparatus.

Control tests on wavelengths of the same length as in the case of the first chamber showed that no signs of reception could be detected once the lower brass edge of the bonnet was everywhere submerged in the mercury. Reception took place the instant even the slightest chink was formed anywhere. A round hole, in the chamber wall, especially made for the purpose of these tests, did not allow any leakage of electric energy to the outside, provided the area of the hole did not exceed 1.5 to 2 cm^2. Reception, though weak, appeared only when a piece of wire was introduced through the hole into the chamber.

The series of experiments described above convinced us that the screening of the chamber constructed by us did in fact electromagnetically isolate the experimenter placed within the chamber from external sources, and also that it did not allow the leakage to the outside of any electromagnetic energy radiating from a brain (if any). To make still more sure, we calculated the screening capacity of our chamber.

It is known from electrical theory that an electromagnetic field penetrates into the conducting medium to a depth approximately equal to one wavelength, and that this wavelength has to be calculated for the given conducting medium.

As is known, the wavelength (λ) is related to the speed of diffusion v in the following way,

(1) $\lambda = vT$,

where T is the oscillation period. The speed of diffusion, in its turn, depends on the physical constants of the given medium, and this interdependence is given by the following equation (in the case of a plane wave)

(2) $v = 1/\sqrt{(\gamma\mu T)}$.

In this equation γ is the specific inductive capacity of the medium, μ its magnetic permeability, T, as above, the period of oscillation.

Introducing the expression of the speed of diffusion into equation (1) we have

(3) $\lambda = \sqrt{(T/\gamma\mu)} = \sqrt{(S/\mu f)},$

where S is the electric modulus of the medium in CGS units of the system, and f frequency of oscillation.

In this way, then, the wavelength in the conductor can be calculated and one can assume that the depth of penetration of the electromagnetic field into its conductor will be the same. Calculating by formula (3) the depth of penetration of a current into iron (taking $\gamma = 10^{-4}$ and $\mu = 10^3$) we can see that for the frequency of oscillation $f = 3 \times 10^9$ c/sec. (the length of the wave in air is equal to 1 m) the depth of penetration of the current is expressed in terms of ten thousandths of a millimetre. For the wavelength in air to equal 100 m we get a depth of penetration equal to 0.018 mm, and for a wave of 10,000 m approximately 0.2 mm.

Thus, the calculation shows that the whole range of wavelengths, beginning with 1 mm and ending with 1 km, will be effectively screened out by our iron chamber. Experimental verification of this may be obtained in everyday radio practice.

It is necessary to be careful concerning waves shorter than ultra-violet. It is known that hard X-rays penetrate metal. It follows that formula (3) can be applied only up to a certain frequency. The experiments of certain scientists (Glagolev, Arkadiev, and others) show that if the wavelength becomes comparable to the size of the molecules then the diffusion phenomena of the electromagnetic field differ from the diffusion phenomena under usual conditions.

Similarly, formula (3) cannot be applied in the case of static fields since it follows from this formula that when $f = 0$, the depth of penetration must be equal to infinity.

To sum up, one must conclude that if, when using a perfect screening chamber, no telepathic effect is found, then the energy producing the telepathic effect must be considered as an electromagnetic one with a frequency lower than that of X-rays. If, on the other hand, a telepathic effect is produced under such circumstances, and if one takes such energy to be of an electromagnetic nature, then one should look for the energy in the frequency range corresponding to shorter wavelengths than the soft X-rays, or with a wavelength measurable in kilometres.

Our second problem was to find suitable experimental subjects—telepathic percipients. Attempts to secure such subjects were made at the beginning of our experiments (1932). Most of our subjects were taken from among the patients of Dr. A. V. Dubrovsky, were aged from 19 to 40 years, were suffering from hysterical and neurasthenic disturbances and could easily be sent into hypnotic trance, usually reaching the somnambulistic stage. In the course of these tests we selected those subjects that seemed most suitable for experiments on mental suggestion. Two hysterics, aged 35, E. M. Ivanova and K. G. Fedorova (both female) were selected.

Meanwhile, all the experimenters were gaining experience and mastery of the practice of mental suggestion (telepathic induction) which requires above all the ability to concentrate in a most intensive manner on the "telepatheme," which is by no means easy and very tiring. I. F. Tomashevsky, A. V. Dubrovsky and, in some instances, L. L. Vasiliev, the director of the project, served as agents or senders.

Finally, the last really important (and also the most difficult) task of all was to find methods of mental suggestion which would achieve more perfectly than methods previously employed the aims of the planned psychophysical experiments.

It should be noted that it is possible to find, in the literature of the subject, descriptions of a number of different methods of mental suggestion, which fall into three classes according to the nature of the transmitted "telepathy": motor, sensory and hypnogenic.

At first we selected mental suggestion of simple acts of movement, and mental suggestion of visual images and sensations; thereafter of mentally waking up and inducing sleep. These methods yielded, in many instances, unmistakably positive results which established the existence of mental suggestion as a fact. But the first two methods proved less suitable for purposes of psychophysical experimentation. Experience taught that they either required a considerable period of time, or yield only "indeterminate" results which made them useless for the purpose of providing the bulk of quantitative data essential for statistical calculation, which is of fundamental importance in experiments on mental suggestion. This was our principal reason for selecting from the methods enumerated above the procedure of mental alerting and lulling, or suggestion to wake up and go to sleep; a method which, in the course of our investigations, gave us the most stable positive results.

4

Mental Suggestion of Motor Acts

We will not here provide a detailed account of all the information concerning influencing bodily movements by mental suggestion which is to be found in the literature. We will, however, mention in the first instance the investigations of Dr. Joire[1] because they are well known, and because they served as the starting point of our work.

In his experiments, which were carried out at the Faculty of Medicine of the University of Lille (France), Joire blindfolded the subject and then suggested that he should perform some movement, stipulated in advance, for example to raise his left arm or hand or his right leg, to cross his arms over his chest, to raise his left arm alongside the body and bend it, to walk in a specified direction, to approach one of the persons present, etc. The experimental conditions were as follows: medical students were employed in the experiments. One of these stood in the middle of a large room, his eyes covered by a cotton-wool padded black bandage. The experimenter (Joire) induced in the subject a state of passivity, i.e. attempted to remove by means of suggestion all extraneous thoughts, and stood either in front of him or behind him at a distance of 3 to 4 metres. Those present were asked to keep silent, and to make neither verbal nor mental counter-suggestions.

The experiments were carried out in the evening in full light. The experimental subject was instructed not to make any voluntary and deliberate movements, but at the same time not to resist moving any parts of his body if he felt moved by an impulse to do so. In the first experiment with any given subject not even that instruction was given. The desired movement was written down previously on a piece of paper. The experimenter endeavoured mentally to divide this planned movement into a series of successive contractions of different component groups of muscles, while fixating the subject with his eyes, and attempting, as vividly as possible, first to imagine such a series of muscular contractions and then to induce the subject to carry them out. According to Joire, a considerable effort of will on the part of the experimenter is necessary, and this must be sustained throughout the entire test.

The results claimed were positive. Some of the subjects (even some who attended for the first time and were given no verbal instructions as to what was required of them) often faultlessly made precisely those movements that were suggested to them. According to Joire it was characteristic of this type of phenomenon that the time interval between the beginning of mental suggestion and the beginning of the suggested movement was 10 to 20 seconds, but that the movement was executed very slowly. After the movement had been performed the bandage was removed from the subject's eyes and he was questioned as to his experiences. All the reports of all the subjects were similar: after the bandage was placed over their eyes and after manipulation by the experimenter, there arose a feeling of isolation from the surroundings, a lowering of the critical consciousness, a numbness pervading the whole body, sometimes a feeling of tingling. This state was followed by an influence of "an alien impulse" (l'influence d'une impulsion étrangère) in a definitely specifiable group of muscles, the subjects succumbed to this influence almost unconsciously, and then carried out the suggested movement.[2]

In 1926, in one of the hospitals of Leningrad, these Joire experiments were repeated by a team composed of the medical hypnotist Dr. Finne, my colleague V. A. Podierny and myself. A chronic patient, Kuzmina, aged 29, was our experimental subject. She was undergoing hypnotherapy in respect of a long-standing hysterical paralysis of the left side. Once under hypnosis the patient remained immobile and her musculature flaccid during the whole session, after which there were signs of neuromuscular hypertension, taking specific forms. This phenomenon had already been noted by J.-M. Charcot and his school, and also by V. M. Bekhterev.[3] Our patient, when in the hypnotised state, displayed a lowered receptivity to verbal suggestion (this being characteristic of hypnotic states of this type). However, under the influence of very persistent repetition of verbal suggestion the hypnotised patient regained her capacity for voluntary movement of the paralysed extremities. A similar effect could be produced by mental suggestion. It should be particularly noted that, although this patient under hypnosis showed lowered receptivity to verbal suggestion amounting at times to negativism, she gave every indication of being intensely receptive to mental suggestion of motor acts.

The experiments were carried out in the morning in a separate small ward from which all furniture was removed except a stool, and a bed for the patient which stood in the middle of the room. Dr. Finne would put the subject into a hypnotic state by means of verbal suggestion. One of the experimenters sat behind the patient's head, at a distance of about 2 m. Either the experimenter, or someone else who was present, wrote on a slip of paper the mental suggestion to be given (often this was not shown to the others present), and the experimenter proceeded with the experiment following the technique of Joire. Especial care was taken both by the experimenter and all those present to avoid unconscious whispering, thus giving verbal information concerning the nature of the assigned task: such whispering may occur spontaneously and involuntarily and would, of course, vitiate the results.

Once the subject was in a state of deep hypnosis (with posthypnotic amnesia but without the instruction of rapport confined to the hypnotist), her eyes being tightly closed or more frequently tightly bandaged, she would usually make the movement suggested to her without any unnecessary movements. When asked by the hypnotist why she had made this particular movement she would often reply, "I was told to do so by so-and-so" (Finne, Vasiliev, etc.) nearly always correctly naming the agent or sender. Below are given without exception all the memoranda that were made concerning experiments conducted with this subject, positive as well as negative, in chronological order.

Experiment 1. 21/8/1926. Task: to stretch both arms outward. The task was set by Dr. Finne who also acted as sender; he was sitting on a stool behind the subject's head at a distance of 1 m. Within 1.5–2 minutes after the beginning of suggestion convulsive movements of the whole left arm (the affected one) were noted, which movements gradually brought the arm into the suggested position. The right arm, after making several indefinite movements, remained in its previous position. The task was considered to have been only partly accomplished.

Experiment 2. 25/8/1926. Task: to cause, by means of mental suggestion, in the subjects right arm the effects of neuro-muscular hyper-stimulation (after Charcot)[4] in the following sequence: elbow (ulnar) nerve—median nerve—radial nerve. The task was set by Dr. Finne who was also the sender and sat behind the patient's head at a distance of 1 m.

I (a). Suggestion: hyperstimulation of the (elbow) ulnar nerve *(n. ulnaris)*.

I (b). Response: correct, on right arm, within 2 minutes, i.e. the hand takes up a position characteristic for stimulation of the ulnar nerve.

II (a). Suggestion: stimulation of the median nerve *(n. medianus)*.

II (b). Response: the position of the right hand resembles the position characteristic for stimulation of the ulnar nerve. Position assumed within 1.5 minutes. Task not fulfilled.

III (a). Suggestion: hyperstimulation of the radial nerve *(n. radialis)*.

III (b). Response: carried out correctly, but on the left arm, the right one remaining immobile (within 2 mins.).

Experiment 3. Same date as above. Task: Cross the arms over the chest. Set by Vasiliev; the sender, Finne, in same position as in experiment 2. Within 1/2 minute after the beginning of suggestion the subject starts to make movements with the right arm in the following sequence: (a) she places her right arm on her chest in the direction towards her left shoulder but then (b) she brings her hand to her mouth and moves it as though wiping her lips with her fist; (c) she rubs her forehead and nose with the back part of the hand; (d) puts her hand under her head.

Dr. Finne asks the subject: "What do you want to do?" The subject replies: "To make the sign of the cross on my face." Question by Dr. Finne: "Why do you want to do that?" Reply: "Prof. Vasiliev made me do it."

Experiment 4. Same date. Task: to sit up and open her eyes. Set by Dr. Finne; sender, L. L. Vasiliev, sitting behind the head of the subject at 1 m distance. Within half a minute from the onset of suggestion the subject begins to make restless movements of the head and then moves in the following sequence: (a) turns head to one side; (b) makes visible effort to raise head; (c) raises head which falls back on cushion; (d) raises head again which falls back once more; (e) raises whole body and falls back inertly; (f) puts her right (healthy) hand behind her head, helping herself up by thus pushing herself up, sits up and remains sitting up with closed eyes.

Question by Dr. Finne: "What did Prof. Vasiliev tell you to do?" Reply: "He told me to sit up, put my hand behind my head (not suggested) and see who is present." About five minutes had elapsed from the beginning of the suggestion to carry out this task.

I must add that in every case the task was written down on a piece of paper behind the subject's head immediately before the beginning of the experiment and shown in silence to four persons present at the sitting (Finne, Vasiliev, Podierny and to one of the persons invited to attend the experiment).

Experiment 5. 1/9/1926. Prof. Vasiliev, without prior notice to those present, suddenly raises his right leg, mentally ordering the subject to do likewise. The subject, almost instantly after the beginning of the suggestion bends her right leg, then raises the lower part of her leg. Question by Finne: "Who told you to do that?" Subject: "It was Prof. Vasiliev's order."

Experiment 6. 8/9/1926. Task: "Show your tongue." Dr. Finne is sending the mental suggestion, sitting on a stool on the left hand side of the subject, looking straight into her face. Half a minute after the onset of suggestion the subject makes facial movements mimicking a smile. She then raises her right arm, attempts to open her mouth with the fingers of her right hand. Question by Dr. Finne: "What are you supposed to do?" Subject's reply: "Blow a kiss!"

Experiment 7. 11/9/1926. Dr. Finne sitting behind the patient, sends out the mental suggestion that he is stimulating the ulnar nerve of her left arm, then the median and finally the radial nerve of the same side. The behaviour sequence is as follows: (a) the patient raises her left arm. The hand assumes a position similar to that which results from mechanical stimulation (after Charcot) of this nerve. Then the arm falls back: (b) the patient again raises her left arm to her face, the hand being in a position characteristic for the effects of stimulation of the median nerve (after Charcot); (c) nervous movements of the right arm. The right hand assumes the position characteristic of stimulation of the radial nerve (after Charcot). The left remains immobile.

Experiment 8. 15/9/1926. Task: "Place both hands under your head." Her left (partially paralysed) hand twitches. She grasps her left hand with her right (sound) one, places both hands behind her head with a sigh of relief and remains in that position. Question by Finne: "What did you do?" Reply: "I placed my hands behind my head. Prof. Vasiliev said I should." Time taken: 4 minutes.

Experiment 9. Same date. Task: "Sit up and open your eyes." Set by Dr. Panov, invited to attend the experiment. Hypnotic sender: Dr. Finne. For five minutes a number of chaotic movements of right and left arms. After a lapse of six minutes from the beginning of the experiment Dr. Finne, the sender, asks: "What were you asked to do?" Subject answers: "I didn't feel anything."

Experiment 10. 25/9/1926. Task set by Vasiliev: "Bend the right leg at the knee." Sender Vasiliev, standing 2 m behind the subject's head. The task is carried out by the subject without superfluous movements, after a lapse of 3.5 minutes. Head, arm and upper portion of body remain in state of complete relaxation.

Experiment 11. Same date. Vasiliev, without any preliminary preparation and without notice to those present, stands behind the subject's head at a distance of 2 m. He raises his right arm, mentally instructing the subject to do likewise. After 2.5 minutes the subject complies in the following sequence: (a) raises both arms a little; raises left arm by taking her left hand in her right hand; (b) lets both arms drop down to her side; (c) with an impulsive quick movement raises her right arm upward; the left remaining at her side. To the question by Dr. Finne "Why did you raise your hand?" the subject replies "Prof. Vasiliev spoke to me about it."

Experiment 12. 6/10/1926.[5] Task set by Prof. A. A. Kuliabko, invited to watch the experiment: "Scratch your left cheek and the bridge of your nose." Sender: Prof. Kuliabko, sitting behind the subject's head. During the experiment the sender raises his right arm and rubs his left cheek. The subject bends her right leg. Raises her right arm to her left cheek and her lips. Scratches right cheek with the same hand. Question by Dr. Finne: "What are you doing?" Subject's answer: "The right side of my face is itching."—"Who spoke to you?"—"Not you."—"Who was it then?"—"Prof. Kulbashov." (The subject had on that occasion met Prof. Kuliabko for the first time; there were 12 people in all present at the experiment.)—"What did he ask you to do?"—"He made the right side of my face itch most horribly."

Experiment 13. Same date. Task: "Open your eyes." Task set by Dr. Finne who also acts as sender, sitting behind the subject's head. After a lapse of 2.5 minutes from the beginning of suggestion the subject several times touches her cheeks, temple and forehead with her left hand and then fully opens her eyes. Question by Finne: "What am I telling you to do?"—reply: "To open my eyes."

Experiment 14. Same date. Task set by Dr. G. V. Reitz, invited to watch the experiment: to make the sign of the cross with her right hand. During the experiment the sender frequently raises his right hand to his forehead, trying to force the subject to do likewise. After 25 seconds the right hand, previously immobile on the bed, assumes the position which the sender is trying to induce. The subject then makes a number of different movements with her right and left arms. She sighs. After 4 mins. 45 secs. she raises her right hand with a visible effort, the fingers bent in accordance with the suggested assignment; convulsive movements of the upper part of the body. Question by Dr. Finne: "What are you doing?"—"My arm is very tired."—"What did you want to do?"—"To straighten the fingers of my right hand, but it was stiff and painful in all the joints."

Experiment 15. Same date. Task set by Dr. Schreiber, invited to the experiment: "Sit up." He acts as sender, sitting on a stool behind the subject's head. Subject shrugs her shoulders. Slightly raises both arms, puts them out, pulls herself up and sits up without opening her eyes (carried out in 2.5 minutes).

An analysis of the above records shows that out of 19 tasks in 15 experiments 10 were carried out quite correctly, 6 may be considered as only partially correct and only 3 yielded either no results or a false response. It is of interest that the subject not only carried out the suggestion of those who usually experimented with her (Finne, Vasiliev) but also those of casual visitors whom she had met for the first time (Kuliabko, Schreiber, Reitz, Panov).

It is a pity that subjects as suitable for experiments in mental suggestion as Kuzmina are rare. At the beginning of our new series of joint investigations in 1932 at the Institute for Brain Research we tried to find new subjects suitable for experiments of this type. 12 hysterical and psychasthenic patients were tested, selected from one of the psycho-neurosis departments. Of these only two proved amenable to the Joire method: Gr . . . na and B . . . na, both hysterics.

Subjects No. 7, Fe . . . va and No. 8, Iv . . . a, who gave no positive results in experiments of mental suggestion of motor actions proved, however, to be highly suitable for suggestions of sleeping and waking. This will be dealt with in Chapters 6 and 7.

From Table 1 it appears that, whereas the experiments with subjects 1 and 2 gave positive results on most occasions, experiments on the remaining subjects were all virtually unsuccessful. By way of illustration I will mention a few of the experiments with the subject Gr . . . na, carried out in my presence by

Dr. Dubrovsky. These experiments did not differ greatly, either as regards experimental set-up or results from those described above with Kuzmina as subject.

Table 1

Serial No.	Name	Total No. of Experiments	OUT OF THE EXPERIMENTS		
			Successful	Partial Success	Unsuccessful
1	Gr...na	18	11	3	4
2	B...na	9	6	1	2
3	Br...i	5	1	0	4
4	Le...va	2	0	2	0
5	S...va N.	2	0	1	1
6	Go...va	7	0	1	6
7	Fe...va	1	0	0	1
8	Iv...va	1	0	0	1
9	Ko...va	1	0	0	1
10	De...va	1	0	0	1
11	S...va M.	3	0	0	3
12	No...aia	4	0	0	4

Results of experiments in mental suggestion of acts of movements by the Joire method.

Experiment 1. 2/6/1932. The subject, Gr . . . na, is lying on a couch in hypnotic trance, the sender, Dr. Dubrovsky, is standing 1 m behind her (time 9:42 p.m.). Task: "Sit up on the couch, putting your feet on the floor down the left side of the couch." Response: subject mutters something. At 9:49 p.m. she makes some efforts to raise herself. At 10 p.m. she sits up and puts her legs down the right hand side of the couch.

Experiment 2. 16/8/1932. Subject: Gr . . . na. Conditions same as previous experiment. Sender Dr. Dubrovsky. Task 1 (10:53 p.m.) "Raise your left arm." Response (10:55): the subject's breathing becomes laboured then (11:05) she slowly raises her left arm without unnecessary movement. Task 2 (11:07) "Raise your right arm." Response (11:07) instantaneous. Subject raises right arm and keeps it up. Task 3 (11:10) "Lower your raised arm." (11:11) the subject lowers her arm.

In 1937 we resumed our investigation of motor acts. A number of subjects were tested by Joire's method but only two proved more or less suitable. L. X.— a perfectly healthy girl of 28, a pianist; and I. M., a woman 33 years of age, suffering from severe hysterical disorders, who had been for a long time in a Leningrad psychiatric clinic, and was under treatment by means of hypnotherapy.

Three experiments were made with L. X. At first the experiments were carried out without any prior instructions and the possibility of the existence of mental suggestion was not mentioned. The subject who had been put into a "passive state," stood with her eyes securely bandaged in the centre of the room. The experimenter (Vasiliev) stood in front of the subject at a distance of 3 to 4 m. Only one other person present in the room. Time of experiment about 10 p.m.

Experiment 1. 12/3/1937. Task: raise the left arm. Before the onset of suggestion the subject stands still, both her arms hanging down alongside her body. Within 30–40 seconds from the beginning of mental suggestion the subject's left arm begins to move, raises itself slightly, then stops. When questioned the subject said she felt nothing.

Experiment 2. Same date. Task: "Approach the experimenter." Within 1 to 1 1/2 minutes the subject begins to lean forward and takes a few steps in the direction of the experimenter. When questioned she said she felt pulled forwards.

Experiment 3. Same date. Task: "Rock backwards and sit down on the chair behind you." After about 2 minutes from the start of mental suggestion the subject throws her head back, rocks about but does not carry out the assignment. When questioned she says that she became sleepy and numb over her whole body.

It was not possible to continue experiments with this subject as she was too busy.

Systematic experiments were carried out with the second subject, I. M., who came to the laboratory for treatment by means of hypnotherapy. The usual procedure was as follows. The subject, sitting in a comfortable chair, was put under hypnosis by means of mental suggestion. The trance was not deep, there was no exclusive rapport with the experimenter, and post-hypnotic hypnosis was only partial. After a number of verbal suggestions of a psychotherapeutic nature the experimenter (Vasiliev) moved away from the subject to a distance of 4 to 5 m, asking her to sleep quietly until he returned. Sitting down at her side but facing her the experimenter, after a lapse of a few minutes, proceeded with the experiments of mental suggestion without giving any verbal (spoken) instructions. After completion of the experiment the subject was brought out of her hypnotic condition by means of verbal suggestion and, when awake, she was questioned. This questioning usually showed that the patient did not remember the tasks suggested mentally, although she was aware of them in the hypnotised state.

Here are the brief reports of all the experiments, without exception, carried out with the subject I. M. in the summer, in the evening.

Experiment 1. 19/6/1937. Task: "Raise the left arm." After 10 to 15 seconds the subject vehemently raises the left arm above the arm of the chair on which it had previously been resting for a long time. The arm stays for some time in the raised position and is then lowered.

Experiment 2. Same date. Task: "Raise your left leg." After 10 to 15 seconds the subject impetuously puts forward the left leg without taking it off the floor. No other movements were noticed.

Experiment 3. Same date. Task: "Raise your right leg." The subject leans forward and tries to get up, but does not succeed in doing so. She is induced by means of verbal suggestion to resume her former quiet position.

Experiment 4. Same date. Task: "Raise your right arm." After 10 to 20 seconds the subject, by a sudden jerky movement, takes her right arm off the arm of the chair and holds it up in the air for some time. Then does the same with the left arm (not suggested).

Experiment 5. 26/6/1937. Task: "Raise your right arm" (previously lying on her lap). After 5 to 10 seconds the subject by a jerky movement slightly raises her right arm and then makes a similar movement also with the left arm, but more weakly and slowly, as though undecidedly.

Experiment 6. Same date. Task: "Raise your left leg." The suggestion is made for several minutes but the subject does not respond. She makes no other movement either.

Experiment 7. Same date. Several minutes before the experiment the subject had spontaneously put the fingers of both hands together, folded her hands and put them in her lap. Task: "Separate your hands." After 10 to 12 seconds the subject, as though overcoming some resistance, carries out the instruction by separating her hands and putting them alongside her body.

Experiment 8. Same date. Task: "Raise your right arm." The mental suggestion lasted several minutes but met with no response. Eventually the subject stretched out her left hand and made her hand into a fist (not suggested).

Experiment 9. Same date. Task: "Stand up." Mental suggestion lasting for 2 minutes without effect. The experimenter then approached the subject, put his hand in which there was an unlighted cigarette over her head and asked: "What am I holding over your head?" The answer: "A bunch of flowers." Thereupon the subject suddenly stands up and takes a few steps. This may possibly have been a delayed compliance with the previous suggestion "stand up."

Experiment 10. 1/8/1937. Task: "Raise your left arm." Within 10 or 20 seconds the subject raises both arms and makes gestures as though chasing away some flies (none in the room).

Experiment 11. Same date. The subject is squeezing the arm of her chair with her left hand. Task: "Let go the arm of the chair and raise your right arm." The task was carried out in 5 or 6 seconds. The arm was raised, left in the air as if frozen there, and then slowly lowered again.

Experiment 12. Same date. Task: "Raise your left arm." (This arm was hidden by her body from the experimenter who was sitting at the subject's other side.) The task was carried out almost instantaneously.

Experiment 13. Same date. Task: "Raise your left arm" (same task). The mental suggestion was continued for 2 minutes but the subject did not comply.

Thus, out of 13 tasks, 6 may be considered to have been carried out with complete accuracy, 3 leave room for various doubts, and 4 were not carried out at all. After the summer vacation hypnotic sittings were resumed in October. A further 10 experiments were conducted which were similar, both as regards procedure and results to the ones described above. Some of the experiments in this second series were carried out in the presence of G. U. Belitzky.

It should be noted, however, that prolonged repetition of experiments of this type with one and the same subject renders such demonstrations less convincing on every occasion. Whereas during the initial experiments the hypnotised subjects are passive, almost immobile, in later experiments, when motor acts are suggested to them again, the same subjects become ever more restless. They make spontaneous movements that are not suggested to them with ever increasing frequency, and this naturally reduces the value of the experiments. It seems as if the hypnotised subject, quite conscious during hypnosis of the voluntary movements such as sitting up, raising arms or legs etc., somehow intends to make them even though he does not, on waking, remember having made them. Thus the experimenter accidentally, so to speak, trains his hypnotised experimental subjects to perform given movements. These remarks apply even more to experiments carried out by the Joire method on unhypnotised subjects who realise what the experimenter expects of them.

In view of these considerations we decided that it would be necessary to start a series of experiments that would overcome these difficulties. The essence of these experiments lay in trying to induce subjects, by means of mental suggestion, to carry out movements that they do not notice even in the waking state. Among such movements is the unconscious swaying of the body when one is standing still. Such body sway can be measured by means of a cephalograph, but this instrument was not suitable for our work. Its use requires the placing of a ring on the subject's head which is connected to levers which transmit the swaying movements to a recording capsule. Not only may the placing of such a ring on the subject's head hamper the subject, but it may also suggest to him undesirable guesses concerning the nature of the experiments.

It is for this reason that we decided to forego the use of the cephalograph and adopt instead, for the purpose of measuring subconscious body sway, the use of a pneumatic platform, designed by A. I. Bronstein. We will now describe Dr. Bronstein's methods as applied by us.

A wooden equilateral triangle is placed on the floor. The triangle has three recesses, one in each corner. In two of the recesses we placed a small wooden cube, in the third a thick-walled rubber squirt. Another similar triangle was placed on top of the first, and the subject stood on this platform for the purpose of the experiment. A thick-walled rubber tube was connected at one end with the rubber squirt, and at the other to a hermetically sealed container, connected by means of another tube with a registering device the point of which touched the soot-covered surface of a slowly revolving kymograph. Air could be pumped into the rubber squirt and air-transmission system by means of a bicycle pump.[6]

The experiments were carried out as follows. The subject was standing on the platform facing the wall, with his eyes closed, in Joire's "passive state," and was requested to stand still during the entire experiment. He was so placed on the platform as to have the corner with the rubber squirt in front of him. In this position his slightest backward swaying movement, even if not detectable by eye, caused a lowering of the water level, and hence a lowering of the curve which was traced on the revolving drum of the kymograph. A slight forward swaying resulted in a raising of the lever and the curve. It is known that every standing person is subject to a certain amount of forward and backward sway, and the kymograph tracing for any given person has a characteristic pattern. The more he sways, the greater the range and the sharper the peaks of the curve. In our experiments the kymograph and the controller working it (Dr. G. U. Belitzky) stood on the right of the subject, at a distance of a few metres, and were concealed from the subject by a curtain. The experimenter (Vasiliev) was on a stool behind the subject at a distance of 2 to 3 metres.

In the first instance we tested our set-up on three subjects who happened to be available—students and laboratory assistants, from 18 to 20. At the beginning of the experiment the subject's characteristic spontaneous swaying pattern was recorded for some length of time, without there being any suggestion of any kind. A mark was then made on the kymograph ribbon, and the experimenter attempted, by means of wordless mental suggestion, to compel the subject to sway backwards and forwards, or to intensify the ordinary forward and backward swaying. This suggestion was kept up for about a minute; the experimenter then stopped, and marked the record again.[7] These experiments gave no noticeable results: the recorded curve was the same as before, during, and after the period of suggestion. The results were, however, quite different when the suggestions were made verbally: "It is hard for you to stand still. You feel as if you were being pulled forwards (or backwards); you are falling, you are falling . . ." The curve, after such suggestions, changed: while verbal suggestion was given, larger, more spiky, fluctuations were recorded, providing evidence for greater amplitude of swaying. The same thing happened when the subject was asked to repeat quickly: "I am falling backwards, I am falling backwards . . ." etc. In the case of auto-suggestion the range of fluctuations increased markedly, but not identically in the case of all three subjects, but depending presumably on their relative susceptibility to suggestion and auto-suggestion.

The number of subjects was then increased, which provided us with the possibility of distinguishing between the various types of so-called "ideomotor" reactions. In the case of some subjects, after verbal suggestion to fall backwards, the curve immediately (Fig. 6Б) or gradually (Fig. 6A), began to fall, followed by a sharp increase corresponding to the extent of sway, and providing a clear picture of the subject's so-called ideomotor reactions. In the case of other subjects, the characteristics of the curves did not change at all (Fig. 6B) in response to suggestion: again, in other cases, we noted reactions contrary to the suggestion: the swaying decreased in response to an instruction to sway more (Fig. 6Г), thus indicating that these subjects displayed a negative reaction to the suggestion, inhibiting even such swaying movements as they had made spontaneously prior to the onset of suggestion.[8]

I. K. Spirtov was the first among Bekhterev's collaborators in the study of ideomotor acts to make experiments of this type. An American, Jacobson,[9] was the first to establish that the imagined picture of a movement or of a visual image is accompanied by a series of rhythmic biocurrents in muscles which create the imagined movements. There is now a big literature on ideomotor acts. M. S. Bikhkov has presented a doctoral thesis on this subject, devoted to the simultaneous registration of cortical rhythms and the skeletal muscular rhythms while the subject is performing an ideomotor act.[10]

Having thus satisfied ourselves of the adequate sensitivity of our technique and apparatus we then proceeded to similar experiments with the subject I. M., already known to readers, who had given us reasonable results of response to mental suggestion when we employed the Joire method. As was to be expected, the standing position of this subject was far less firm and stable than that of previous subjects, none of whom was subject to any mental disorders: the curve level indicated very pronounced swaying even when no suggestions were made to this subject, particularly at the beginning of every experiment. Before every experiment a "passive state" (in Joire's sense) was induced in her, and she was then requested to mount the platform and to stand still during the experiment: that was the full extent of the verbal instructions she was given. With this subject experiments of *mental* suggestion only were carried out. All experimental conditions and the contents of mental suggestion were the same as those in the case of the three healthy subjects. The result, however, was strikingly different—it was distinctly positive.[11]

At the beginning of the experiment (prior to the onset of suggestion) it was possible to discern fluctuations considerable in range, but slow and smooth, indicating the unsteady standing position characteristic of nervous patients. After some time the experimenter proceeded to make mental suggestions, e.g. "sway backwards, sway backwards." This triggered off a number of sharp swaying movements, successive swings almost merging into one another. After the termination of mental suggestion the previous more stable pattern re-established itself. During a second mental suggestion to the same effect the same reaction took place as before, but was somewhat less extreme: we noted once more a series of rapid fluctuations, but not quite so extensive as in response to the first suggestion. During the third bout of mental suggestions, however, the subject manifested the most extreme and rapid swayings of all, indicating that she had, in the course of the experiment, so to speak lost her balance.

We were thus in a position to conclude that, in the above experiment the characteristics and intensity of the involuntary swaying movements of the subject definitely changed three times in succession, at times coinciding with the periods of mental suggestion.

Questioning the subject after the experiment showed that she herself had not noticed the increased bouts of swaying which were kymographically registered during mental suggestion. To the question: "What did you feel during the experiment?" she would reply: "I did not feel anything." All signals that might have given the subject any clue as to the beginning and ending of mental suggestion were avoided. The subject had no knowledge of the purpose of the apparatus or what the experimenter was recording or what she was supposed to do. It would seem that all possible methodological errors were avoided in these experiments, and that the results have to be attributed to the mental reaction of mental suggestion.

It should be noted, however, that positive results of this kind were not obtained in all experiments. In a number of cases the mental suggestion was without effect. In other cases unsuccessful experiments alternated with manifestly successful ones.

For example, at the beginning of the kymographic record we saw groups of weaker and stronger unsteady vibrations. The mental suggestion "sway back and forth" changed nothing. A second attempt changed the picture completely: in this instance, while the suggestion was being made, there is an indication of greatly increased fluctuations of the curve, both backward and forward sway being increased equally: in this case the swaying movements were visible to the eye. It should be added that such strong rhythmic fluctuations of the curve as took place during mental suggestion were never noted in the absence of such suggestion.

It is also worth while noting that our best subjects for mentally suggested motor acts, such as Kuzmina, L. X. and I. M. were less suitable for experiments involving mental suggestion of visual images. Conversely, a number of subjects who consistently gave strikingly positive indications of sensitivity to mental suggestion of visual images proved incapable of responding to mental suggestion of motor acts. There would seem to be no detectable positive correlation between these two variables of telepathic ability. We know that the same holds good for ordinary verbal suggestions. Some hypnotised subjects comply more easily with verbal suggestions of visual images, others are more responsive to suggestions of motor activity.

In summarising our results obtained by the Joire method we wish to emphasise three principal findings:—

1. Our data support the possibility of affecting conscious, voluntary movements by mental suggestion.

2. Our data also support the possibility of affecting subconscious involuntary movements by mental suggestion (e.g. body sway while standing).

3. This latter phenomenon was demonstrated only in subjects who were able to respond to mental suggestion in the manner described by Joire.

Of all the experiments on acts of movement known to me, the most convincing ones in their methods and results are those that were carried out by the Dutch researchers Brugmans, Heymans and Weinberg at the Physiological Institute at Groningen.[12] Two rooms were fitted up for these experiments, one room above the other, and connected by means of a window made in the floor of the upper room. This window was glazed, the glass being so thick that even loud noises did not penetrate from one room to the other. Through this window from the darkened upper room the experimenters could watch what was going on in the brightly lit room below where the subject was encased in a cardboard cubicle resembling a cupboard, closed from the top and all four sides. A table stood in front of this cupboard, just under the window. On the table there lay a piece of cardboard resembling a simplified chessboard with 48 large squares, each marked by means of a chess identification. The subject put his hand through an aperture in the cupboard and rested his hand on the chessboard. Thus, whereas the subject could not see the chessboard, the experimenters, while not being able to see the subject himself, were in a position to watch the movements of his hand.

The experiments consisted in an attempt by the experimenters, looking down from above onto the board and the subject's hand, to direct the hand by means of mental suggestion to the intended square. The particular square on which the subject was directed to place his hand was selected, every time, by casting lots. Under these conditions, which apparently excluded all sensory communication between subject and experimenters, one of the subjects, the student van Dam, correctly moved his hand to the intended square 60 times out of 187; this amounts to 31 per cent of correct answers[13] and very greatly exceeds the number of correct guesses that would be expected by chance in accordance with probability theory, which leads to an expectation of 4 out of 187.[14] After the subject had consumed a small amount of alcohol (30 g) the number of successes increased: without alcohol 22 out of 104, after taking alcohol 22 out of 29. These outstanding experiments, exceptional both in their experimental design and also as regards the results obtained were, I regret to say, not repeated by any other investigators.

The demonstrations of V. L. Durov with trained dogs, famous in their day, fall within the scope of a discussion of mental suggestion of acts of movement. He noticed that animals trained to (will-less) obedience often carried out tasks which were merely in their trainers' thoughts. Durov began to use this to put on turns in his circus, e.g. mentally compelling a dog to bark the number of times that he was thinking, or to bring some object determined by thought.

At first these experiments were carried out by Durov himself, or by some one else in his presence. However, in those conditions the trainer was able to signal to the dogs without attracting the notice of those present, for instance by means of a Galton whistle which produces a high ultrasonic note, inaudible to humans but within the ordinary range of hearing of a dog. Moreover, deliberate or unconscious gestures on the part of the trainer or of the persons present at the demonstration could serve as cues for the dogs, giving them appropriate instructions.

V. M. Bekhterev subjected these claims of V. L. Durov to a careful scientific examination. In order to exclude the possibility of the above mentioned sources of error V. M. Bekhterev and his colleagues set up experiments under the following conditions: 1. The owner (Durov) of the dog was absent. 2. The audience, i.e. the public, was also absent. 3. The dog was brought into the room just before the experiment. 4. After having made a mental suggestion the experimenter hid himself from the dog behind the screen or in some other way. 5. The assistant alone supervised the dog without knowing in what the task suggested to the dog consisted, since the task written down by the experimenter was not imparted to the assistant until the end of the experiment.[15]

V. M. Bekhterev, in the articles describing this work, provides the notes, made at the time, of two experiments carried out by his collaborator A. G. Ivanov-Smolensky (experimenter) and P. Flecksor (assistant), which complied with all the conditions laid down by Bekhterev. We here give the notes, made by the experimenter:[16]—

"Durov (the dog's owner) is absent. I am writing down the assignment in room B: the dog is to pick up with its mouth a ball of crumpled paper which is lying under a table in room A. The assistant (who does not know the nature of the task) lets the dog (a fox terrier bitch called Pikki) into room A. He places the dog on a chair, and holds her head. I stand on the threshold of the door, in front of the dog, at a distance of about 1 m. At the first attempt on the part of the dog to free herself I quickly step back into room B, closing the door. According to Dr. Flecksor's report, Pikki, having jumped off the chair, runs to the ball of paper designated by me which is lying under the table, pokes her nose into it, and then runs in turn to two other paper balls, treating them in the same way. (There were 7 paper balls in all, thrown about in various places.)"

Another report runs as follows:—

"I [Ivanov-Smolensky] send the mental suggestion to the dog that it should run from room A into room B, and there to jump onto a chair. The dog rushes into the other room. Dr. Flecksor, who does not know the nature of the task, follows the dog and closes the door behind her. According to his statement Pikki jumps onto a divan standing next to the chair mentally designated by me and scratches the side of the divan with her paws."

V. M. Bekhterev considered these and the other similar experiments in this series remarkable and deserving to be taken account of, irrespective of the various possible interpretations.[17] In his summary he draws a number of inferences from which I will quote as follows:—

"1. Direct effects by means of so-called mental suggestion on the behaviour of animals can easily be elicited in dogs especially trained to obedience.

2. The effect is obtained without any direct contact between the sender or agent and the percipient (the dog); also whilst agent and dog are separated by some barrier preventing visual cues, such as bandaging the sender's eyes, placing a *wooden* or *metal* [my italics—L. L. V.] screen between himself and the dog, or wearing paraffin covered eye shields; the effect may also be obtained where the sender is deprived of the possibility of continuing to give mental suggestion to ensure mental compliance.

3. Direct effects are obtained without any signals which might serve as instructions to the animals in experiments of this sort.

4. The whole question of direct influencing becomes a matter of physiological laboratory experimentation and susceptible to investigation from all points of view in the sense that both the conditions governing the transmission of and the response to mental suggestion can be determined" (pp. 161–2).

In 1925 Prof. Zimmer of the University of Berlin devoted an article to a detailed analysis of these experiments of Bekhterev's. Zimmer, a very learned zoologist, reached the conclusion that, after these experiments, a "mental connection" between man and animal must be considered highly probable. Experiments of response to mental influence in animals appear to this author to be easier than experiments on human subjects, and he attributes special significance to them.

After the death of V. M. Bekhterev, Durov experiments of this type on dogs were, so far as I am aware, not repeated. Parapsychologists, however, remained interested. Many years later an article by Ferriére[18] was published in the Proceedings of the Institut Métapsychique International, in which Bekhterev's experiments were described and evaluated as having yielded positive results. The well known American parapsychologist Rhine[19] also attributes great value to these experiments.

It is not, however, possible to accept these positive evaluations without some reservations. Even the experiments described above, which V. M. Bekhterev particularly singled out, were not without faults. Balls of paper thrown on the floor would naturally attract the attention of a dog as something new and unexpected. Pikki first rushed to the ball "designated" by the experimenter and then towards the other balls thrown all over the floor. We do not know how the balls were placed on the floor; perhaps the "experimental" ball was nearer the animal than the others at the start, and perhaps this is why she went to this one first; the ball was not "taken in the mouth" as was suggested by the sender but rather was sniffed at by the dog.[20] In the second experiment the animal carried out the suggested task incorrectly. She did not jump on the chair as had been suggested but onto the divan. Other more complicated experimental tasks described in the paper under discussion were carried out more accurately by the experimental dogs, but in these experiments conditions that Bekhterev considered essential were not complied with.

A severe critic will be able to find fault with our own experiments in accordance with the Joire method. In the first place, when these were carried out the agent or sender was sitting behind the percipient, at a distance of not more than a few metres. This, in experiments of mental suggestion, is a possible source of error as a result of which sensory or subsensory (sub-threshold) cues[21] might have been given, and this is difficult to eliminate. Secondly, the number of different movements which can be made by the subject under the influence of suggestion, and whilst either standing upright or lying on a bed, is somewhat limited. Consequently the percentage of chance coincidences between movements suggested and movements performed by the subject's own choice independently might be high.

In our view the experiments in which the pneumatic platform was used deserve greater attention. Our subject I. M. regularly reacted to some extent to mental suggestion designed to upset her equilibrium when she was standing. This could have been of advantage in repeating the experiments while screening the agent or sender of the suggestion. In this case, however, we did not succeed in carrying out such an experiment.

After V. M. Bekhterev had indicated that he had used a metal screen in his experiments with dogs (cf. his conclusion 2), this method was adopted by the engineer B. B. Kajinsky in his experiments with trained dogs at the Applied Comparative Psychology Laboratory.[22] The sender was placed in an earthed screening chamber made from sheet roofing iron, and mentally suggested to the dog under investigation the performance of some motor task. The assignment was more or less correctly fulfilled only in those instances where the sender could watch the animal through an opening the size of a human face, made in the chamber. When, however, this window was closed by a shutter made out of the same roofing iron the dog usually failed to carry out the allotted task.

It is apparently on the basis of these experiments that B. B. Kajinsky favoured the conclusion that screening by means of iron impairs the effect of mental influence by the experimenter on the behaviour of the animal, and he inferred from this that, in all probability, mental influencing or suggestion was effected by means of electromagnetic energy generated by the brain of the sender.

It is not possible to deem these experiments as by any means perfect or the conclusion as warranted by them. In the first instance, B. B. Kajinsky's chamber, which had a sizeable window that could be closed by means of a metal shutter, provided no guarantee of trustworthy screening. None of the necessary checks and control experiments were made: these are indeed necessary, as we found out by experience, cf. Chapter 3. Secondly, the experimental setting of these investigations to discover the effects of mental suggestion on dogs did not satisfy the conditions laid down as essential by Bekhterev: the sender himself was watching the behaviour of the dog when the window was open, and also the trainers present at the experiment could unconsciously and involuntarily (and some of them perhaps did it consciously and deliberately) direct the dog's behaviour by means of some signal, by movement, gesture, facial expression, etc.

We will refer to two further attempts to hinder mental influencing by means of metal. These, as we found later, coincided in time with similar attempts made in the '30s. One of these was made by the medical hypnotist, Dr. T. V. Gurstein, in conjunction with the physicist L. A. Vodolazky in 1936.[23] For these experiments a screening chamber was prepared, made of a wooden frame covered with sheet brass 1/2 mm in thickness: the door and ceiling were covered by two layers of thick brass netting, all seams were carefully soldered. The screening properties of the chamber were tested by means of a shortwave transmitter whose output power was 10 units, and a receiver whose sensitivity could be modified by controlled positive feedback.

It was expected that the percipient would respond to mental suggestion only when the screening capacity of the chamber was impaired, i.e. when the door was open, and that he would not respond to suggestion under "conditions of perfect screening," i.e. when the door was closed.

The experimental procedure was as follows: The hypnotist (Dr. Gurstein), together with one of the collaborators, decided upon the nature of the suggestion and the exact time when these suggestions were to be sent to the percipient. The hypnotist did not know whether the chamber would be open or closed during the transmission of any given suggestion. After the percipient, who was seated on a chair in the chamber, had been put into trance the hypnotist together with his collaborator went into another room from which the suggestions were to be "sent." Another collaborator, who did not know the nature of the suggestion, remained in the chamber. His task was to note the time and the reaction of the percipient, and to close and open the door of the chamber at his own discretion.

In our view this latter decision was a flaw in an otherwise well designed experiment: this collaborator, who was in direct contact with the percipient, could be presumed to have the same expectation as the hypnotist, namely that when the chamber door was closed the suggestion would be ineffectual: and he could have conveyed this expectation to the percipient, by means either of ordinary or of mental suggestion, thus affecting the course of the experiment. For another thing, the percipient could not fail to notice whether the door was open or closed. From our own experience we became convinced that in this type of experiment it is more satisfactory to screen the sender and not the percipient who usually possesses a much higher degree of sensitivity and suggestibility.

The tasks suggested were of the same type of motor acts as those in Joire's and our own experiments, already described in this chapter: e.g. "Raise your right arm," "stretch out your left arm," "make your right hand into a fist" and so on.

The number of such experiments was very small; only ten in all, five with an open and five with a closed door. In all five experiments with a closed door the subject failed to respond to the mental suggestion. In four out of five of the open door experiments the subject correctly carried out the experimental task. It is a pity that the number of experiments was so very small—it is quite inadequate for the purpose of inferring anything definite.

At about that time similar experiments were carried out by the physicist S. I. Turligin, a collaborator of Academician P. P. Lazarev's. The experimental set-up was as follows:—[24]

A large room was divided in two by means of a partition in which there was a large round opening. In one part of the room there stood the screening chamber which was completely covered with lead and had a lead pipe let into the front wall. The pipe was covered over with paper so that the sender, who was sitting in the chamber, could not see what was going on outside. In front of the mouth of the pipe, suspended by long thin threads, there hung a lead screen which could be noiselessly moved to one side. Whether this screen remained in place, over the mouth of the pipe (in experiments *with* screening) or was moved (in experiments *without* screening) was unknown both to the sender in the chamber and also to the percipient who was in the other part of the room divided off by the partition. The screen was left in place or moved by the experimenter in charge of the experiment at his own discretion without any previous plan.

Communication between experimenter and sender was by means of light signals: bulbs of different colours—red, blue and green—indicated what the sender was supposed to do, namely to start suggesting, to stop suggesting, or to remain neutral.

The experimental subjects used for these experiments were perfectly normal and healthy. In the opinion of S. I. Turligin subjects suffering from hysterical and other neurotic disturbances were unsuitable for experiments of this type. In these experiments the subject on the opposite side of the partition from the sender's chamber was instructed to stand with his back to the partition at a distance of about 2 metres from the hole. His head was thus in line with the mouth of the pipe. The experimenter explained to him that it would be mentally suggested to him to fall backwards, but that there was nothing frightening about this since there were assistants standing behind him who would always be in time to catch him.[25] These verbal instructions, even without mental suggestion, were sufficient to induce the subjects to fall backwards within 1 or 2 minutes after the command was uttered in a low voice: "We are starting!" When the sender exerted mental suggestion this period was sometimes reduced to between 2 and 10 seconds.

As for the inhibition of the transmission of mental suggestion by means of the lead screen, the author gives the following numerical results which, in our opinion, are not adequate. "The percentage of successful results, i.e. confirming our experimental hypothesis [i.e. that lead will screen out suggestion—L. L. V.] is more than 76. The average reaction time when the tube was open [no screening—L. L. V.] was 0.8 minutes; the average reaction time when the tube was closed [screening—L. L. V.] was 1.3 minutes." Since the author does not give the total number of his experiments it is quite impossible to assess the significance of his figures.

The phenomena described in this chapter may be regarded as being of the lowest stages of mental suggestion. The motor acts suggested to the percipients, whether these be dogs, or sleeping or even waking human subjects, are often hardly noticed by the subject. The mental element hardly comes in. However, even more primitive, purely physiological, reactions elicited at a distance by one person from another are described in the literature.

An interesting example in this respect is the work of the well known hypnotist and university teacher, Sydney Alrutz.[26] The subject was seated in an arm-chair suitable for the experiments, and was put into deep hypnotic trance by the ordinary methods. A black opaque sack was placed over the sleeper's head, his ears were tightly plugged with cotton wool. Oblong wooden boxes were affixed to the arms of the chair in which the subject inserted his arms which were bare to the elbows. The tops of the boxes were covered by sliding screens made of metal, glass and cardboard. In some experiments the subject's arm was enclosed in a glass cylinder.

Alrutz contends that under these experimental conditions the approach of the hand of the hypnotist (or that of some other person present) caused, through the screen of glass or metal, definite changes in skin sensitivity and muscle contraction of the subject's arm. For example, the subject's hand in the glass cylinder is made into a clenched fist; the hypnotist points his finger at a particular point on the skin under which there is a nerve or muscle tendon that stretches out the fingers, and this is sufficient to induce the subject to unclench his fist and stretch out his fingers. Quite generally, according to Alrutz's data, whichever nerve, tendon or muscle was approached by a hand, whether of the hypnotist or any one else, it responded in an anatomically and physiologically appropriate manner, even when experimenter and subject had no idea of the physiologically correct response.

Alrutz concludes that the human hand radiates some sort of energy able to penetrate screens which provide insulation from draft, heat and electrical stimulation, resulting from the approach of a hand. This supposed energy is thought of as physical since it penetrates through some media (metal, glass) but is absorbed by others (e.g. cardboard).

We should not mention these strange phenomena, which call to mind the long-discarded claims of the so-called magnetists, had we not noticed something similar in the experiments with our remarkable subject Kuzmina. Here, for example, is an extract from the report of an experiment conducted on August 28th 1926:—

"At 10:30 Dr. Finne, in the presence of L. L. Vasiliev and V. A. Podierny, puts the patient (Kuzmina) into trance by means of eye gaze and verbal suggestion. The patient responds quickly, going into hypnotic trance, lying on a bed in a horizontal position, face upward, eyes closed.

The phenomena under investigation are those of neuromuscular hyperstimulation (after Charcot) resulting from the approach of the hand of the experimenter to the right (healthy) and the left (partly hysterically paralysed) hand of the subject. Dr. Finne points with the tips of his fingers (which are stretched out and close together) to various points of the subject's bare arm etc., keeping his finger tips at a distance of 1/2 to 2 cm from the skin surface. The following motor reactions are observed: (a) the experimenter's hand approaches the elbow joint *(n. ulnaris)*—reaction: the hand takes up the position characteristic for pressure on this nerve; (b) the experimenter's hand approaches the shoulder from the outside *(n. radialis)*—reaction: the hand takes up the position which is the characteristic response to pressure on this nerve; (c) the experimenter's hand approaches the shoulder towards *n. mediani*—reaction: the hand takes up the position which is the characteristic response to pressure on this nerve.

The same sequence of events followed in the case of the right (healthy) arm. These reactions were perfectly reliable and regular: they could be elicited however many times one wanted, not only by the hypnotist but by any person present. The reaction appeared almost immediately after the approach of the fingers, became stabilised within 10 to 15 seconds, and disappeared almost instantaneously after the experimenter's hand was withdrawn.

The eyes of the subject were closed but not bandaged. Direct visual cues were avoided by the fact that the subject's head was turned in the opposite direction from that of the side on which experiments were being conducted."

It is a pity that we did not, in these experiments, place any screen (metal or glass) between the experimenter's fingers and the subject's skin. We cannot therefore exclude the possibility that we were dealing with ordinary conditioned motor reflexes (possibly sub-threshold ones) in response to the warmth of the fingers of the experimenter, or to the air vibration at the approach of the hand: indeed this is not inconceivable, since during the days preceding the above described experiments Kuzmina had frequently manifested these reflexes of neuromuscular hyperstimulation in response to pressure on the relevant nerves; and in the course of these pressures on the nerves (unconditioned stimuli) these reflexes would frequently coincide in time with heat or air pressure (conditioned stimuli or signals).

On the other hand the control observations made in this same experiment of August 28th 1926 militate against this supposition. We had first tested the effect of the approach of the hands of the hypnotist to the bare skin of the subject, and this had elicited distinct reactions of attraction of the skin towards, and repulsion away from, the hand. Such reactions appeared right away, in the very first test: and consequently it must be inferred that the experimenter's hand acted as an unconditioned stimulus.

Alrutz, in his work already cited,[27] describes an exactly similar observation, but under conditions of screening: "The subject feels his hand getting lighter: this is unquestionably connected with the phenomenon of the attraction of the subject's hand towards that of the experimenter; the subject's hand will rise and follow the hypnotist's hand even when a glass plate is interposed between them. The feeling of weightlessness may be interpreted as being an expression of the desire to push away the hand of the experimenter" (p. 283). The "new rays" supposed by Alrutz to emanate from the experimenter's hand work on the muscles and tendons of the hypnotised subject through the motor nerves. He also observed the effect of a hand on vasomotor nerves.

It is worthwhile comparing these findings of Alrutz with those later obtained in the course of his work by Rudolph Reutler, director of the malaria research laboratory at Rosh Pina, Palestine.[28] He discovered an effect on the part of living organisms at a distance on the automatic movements of exposed inner organs (ovary, digestive tract, etc.) of insects. The best results were obtained using young female grasshoppers *(acrididæ)* when the air temperature was sufficiently high. The head and extremities of the insect were cut off. The peritoneal sac was cut open; the peritoneal nerve chain was removed. The preparation was filled with saline solution (0.6 per cent NaCl). After this the intestines began to contract rhythmically, and when the preparation was put in a Petrie dish with a glass cover[29] they kept up the movement for 10 hours. The observations on the intestinal movements were made through the glass cover of the Petrie dish by means of a binocular low power simple microscope with a magnification of 20. In the presence of the experimenter the intestines contracted rapidly and strongly. The preparation was then left in an empty room for half an hour and no one was in the adjoining room, either. During that time the preparation became quiescent—the intestines contracted in a slow rhythm. But as soon as the experimenter entered the room, and increasingly so when he sat down to observe the preparation through the magnifying glass, the movements gradually grew stronger and more rapid, reaching a maximum at the end of the fourth minute. In order to avoid increasing the temperature of the preparation with his breath the experimenter held one of his hands in front of his nose and mouth and directed his exhalation above the Petrie dish.

The author remarks that it is hardly to be supposed that the rhythmic contracting activity of the intestines should increase so sharply in response to the heat emanated by the observer since the preparation was inside a covered Petrie dish and under a layer of saline solution. Besides, the air temperature of the room was kept at a very high level (28 to 30º). Nevertheless, control experiments were set up in which the bulb of a sensitive thermometer was placed in the Petrie dish instead of the preparation. During a 15 minute period of observation, in which the breath was not diverted from the dish, there was no change in the thermometer reading.

It was noticed that the effect on the intestinal movements depended principally on the contraction of the respiratory muscles of the observer who was sitting at the table looking through the microscope. A still more marked effect appeared when the observer who was sitting by the microscope sharply contracted his leg, arm or masticatory muscles, or when he made rapid finger movements. A sharp increase in intensity and frequency of the movements of the preparation was noticed 15 to 30 seconds after the onset of muscle contraction on the part of the experimenter; the reaction then diminished, but when the experimenter's muscles relaxed again the reaction temporarily increased.

In 1940 I asked V. S. Steblin-Kamensky,[30] an entomology student at Leningrad University, to repeat Reutler's experiments in my laboratory at the Institute for Brain Research. In the winter of 1940 he carried out a series of experiments, using the larvae of Asiatic Locusts and adult cockroaches; in the ensuing summer he used cockchafers. Steblin-Kamensky precisely followed the method of preparing the insects described by Reutler, but the conditions in which the experiments were carried out were less favourable: the air temperature and saline solution were comparatively cold, in winter there was relatively little radiation from the sun, the number of insects available for experimentation was small; all these factors made the work much more difficult. The unfavourable conditions negatively affected the sensitivity and the motor automatisms of the intestines of the insects under investigation: nevertheless in some of the experiments the Reutler phenomena were observed with sufficient clarity. Here are the reports of two experiments:—

Experiment 1. 29/2/1940. "Experimental object: a cockroach. Method of dissecting preparation: following Reutler. With a magnifying glass with a magnification of 4 diameters only a contraction of the large intestine could be seen (the malpig vessels were too small for the given magnification). After half an hour from the preparation of the insect normal peristalsis of the bowels could be observed. Contractions occurred every 5 or 6 seconds and were well marked. The experimenter was seated at a table and observed the preparation through the magnifying glass. The moment he tensed the muscles of his legs which were under the table stronger, more marked and jerky contractions of the preparation's bowel movements were observed. Whilst the muscles of his legs remained tensed peristalsis resumed its initial level, but as soon as the experimenter relaxed his leg muscles, strong jerky contractions of the preparation's intestine were again observed. These phenomena were observed repeatedly over a period of five minutes. The preparation did not react to a contraction of the fingers."

Experiment 2. 15/6/1940. "Experimental object: a cockchafer, half grown female at time of ovulation. Method of dissecting and preparation: after Reutler. After 20 minutes from the beginning of preparing the insect one could see through the magnifying glass (magnification 6) the onset of the peristaltic wave gradually passing from the upper to the lower intestine. The contractions of the bowel occurred at intervals of 5 to 6 seconds. Those of the malpig vessels at intervals of 2 to 3 seconds. Two or three seconds after a peristaltic wave, when the intestines were at rest, the experimenter heaved a deep sigh. At the moment of the sigh there appeared, out of phase, an energetic contraction of the upper part of the intestine, which spread through the middle down to the lower intestine. After three or four seconds the experimenter made a brisk exhalation which again caused a strong peristaltic contraction, and a similar effect was obtained if the experimenter strongly contracted the muscles of his arms and legs. The peristaltic waves similarly, under the same conditions, increased when passing through the malpig vessels (at onset and termination, i.e. tensing and relaxation of the muscles of an extremity). All these repeated reactions were observed in the course of 20 to 30 minutes."

It should be added that positive results were not by any means obtained in all the experiments, particularly those conducted in winter. The experimenter attributes this, not without good reason, to the rather low air temperature. In the 4 experiments carried out during hot summer (temperatures 20 to 25) positive results were obtained far more frequently. In one of the experiments on a cockroach preparation a positive reaction was elicited within 5 or 6 minutes. At a signal from an onlooker (a colleague in the laboratory) the experimenter contracted the muscles of his hand; this immediately caused a responding peristaltic contraction of the intestine.

These experiments are of interest because a physiological detector—the muscles of an insect's intestines—was successfully used to react to the influence emanating from the contraction of human muscle. Is there any such physiological detector that could register the energetic impulses of an active human brain? The discovery of such a detector could inaugurate a new era in the study of the phenomena of biological wireless communication.

It should be noted that, according to the data published by Wachholder and Altenburge[31] the electrical currents of the active human skeletal muscles, when slightly contracted voluntarily, behave in a manner precisely similar to that of the insect intestines in the Reutler experiments: the initial spikes at the beginning of contraction, the "quiescent period" during a prolonged contraction, and another group of spikes at the relaxation of the contraction.

So marked a similarity cannot be a chance coincidence. It is possible to assume that the phenomena observed by Reutler and by Steblin-Kamensky are caused by the stimulating effect of the low frequency electromagnetic field that results from the biological currents of the observer's muscles. It has long been known that when human muscles contract voluntarily a biological current with a periodicity of 40 to 50 cycles per second flows through them (N. E. Vedensky, Piper, etc.). That such biological currents in fact generate electromagnetic fields which spread in space, and can be detected at a distance, is established by the celebrated experiments of Saurbruch and Schumann.[32] In these experiments an insulated metal rod terminating in a disc was inserted into a Faraday cage through a hole in the wall. The end of the rod outside the cage was connected with a string galvanometer *via* a three valve amplifier. The subject inside the chamber brought his naked forearm to within a few centimetres of the metal disc at the end of the rod; this left the string of the galvanometer unaffected. However, under these experimental conditions it was enough for him to move his fingers for the string to register a number of rhythmic oscillations. Such effects were observed, not only in the case of the voluntary contractions of human muscles, but also in that of the convulsive muscle movements of a rabbit that had been poisoned with strychnine.

In 1934 the engineer R. I. Skariatin attempted to repeat these experiments at our laboratory. A three valve amplifier was constructed whose circuits corresponded exactly to the set-up in the work described above. The amplifier worked on valves of the UB-107 type. Either a telephone or a needle galvanometer could be connected with the annode of the last (third) valve. The annode potential was supplied from a separate battery. The outward appearance of the apparatus is depicted in Fig. 7. The screen of the grid of the first valve was earthed through the coating of the Faraday chamber. The grid of the first valve was connected to a brass rod which passed through an ebony plug affixed to an inside wall of the chamber.

By means of this apparatus R. I. Skariatin carried out exploratory experiments which allow one to expect positive results when the apparatus is improved. It would probably be well worth while resuming these experiments, using modern and probably more sensitive radio apparatus (see Chapter 10). The existence of a weak low-frequency field surrounding a stimulated sciatic nerve excised from the body of a frog is apparently definitely established (B. V. Kraiuchin, in the laboratory of Academician A. V. Leontovich, see Note 27, Chapter 2).

5

Mental Suggestion of Visual Images and Sensations

This is the most usual form of experiment in mental suggestion. In the non-Russian parapsychological literature there are a number of monographs devoted to a detailed description of many hundreds of experiments of this type, together with a critical analysis of the results obtained.[1] These monographs refer chiefly to work done in the twenties and thirties of this century, and their publication dates coincide closely with similar experiments carried out over the same years in the Soviet Union, but not published even now.

As has already been mentioned a large amount of Russian work in this field was done by the Commission for the Study of Mental Suggestion, founded by V. M. Bekhterev.[2] In 1924 the members of the Commission submitted to the Second All Russian Congress of Psychoneurology[3] the results of two years' work, embodied in a paper called "A study of the phenomena of intercerebral induction and perception."[4] It seems to me valuable to mention here the principal findings of this joint work, the results of which in certain respects outstripped those obtained by foreign contemporary scientists.

The first series of experiments consisted in mentally suggesting to two percipients drawings prepared in advance. These subjects had previously demonstrated their ability to do automatic writing by means of the so-called planchette (a small wooden board on three legs, two of which end in rollers and the third is a sharpened pencil). The percipient puts his hand on the board and the hand, as a result of subconscious movements, pushes the board over a piece of paper: the pencil draws pictures, writes letters, words and even whole phrases. The percipient may carry on a lively conversation with the sender, without being aware that his hand is drawing at the same time.

In general, three members of the Commission were present during the experiments. One of these was the sender, the other two observed the sender's behaviour and made notes. The percipient was sitting at a table with the planchette, the sender sat a few metres behind her. The experimenter handed the drawing to be suggested to the sender who would then proceed to try to transmit the mental picture to the percipient.

Unfortunately only one report of these experiments has been preserved among my papers: the pictures drawn automatically by the percipient M. A. S., and the notes taken, have not been kept. Here is the report of V. A. Podierny who acted as sender:

Task 1. Mentally transmitted: a cross on top of a square. The subject tries to draw a square and writes something illegible. Task not fulfilled.

Task 2. Transmitted: dial of a watch with two hands. The percipient writes two pages which have nothing to do with the task, then writes out the word "face," finally draws picture of a watch and two hands, but not in same position as in the image transmitted by the sender.

Task 3. Transmitted: triangle with a circle inside it. The subject completes the task at once.

Task 4. Transmitted: a simple pencil drawing of an engine. The percipient carries out the task precisely and goes over the contour of the engine several times.

Task 5. Transmitted: crescent moon on top of a pole. Percipient first draws square, then triangle. Task not fulfilled.

Percipients who have no aptitude for automatic writing do not obtain such successful results as these—in the above experiment three drawings out of five were distinctly reproduced. I tried for many years, but in vain, to find a similar percipient. Only in the middle of the thirties did I have such an opportunity, and then only for one sitting. Let us leave aside for the moment a description of the materials of the Commission, particularly since the experiment now to be described has a bearing on it.

My former student C., a young girl keen on sports, dropped in to see me one day; she had just graduated from the University where she had specialised in physiology. I asked her to let me test her suggestibility. We started with the pendulum test, as a result of which it emerged that she possessed unusually marked ideomotor reactions.[5] We then discovered that she had the ability to produce automatic script. Without interrupting a lively conversation, whilst holding a pencil in her hand, C. unconsciously wrote on a piece of paper, placed under the pencil, separate words and phrases which were partly legible. Fig. 8 shows her first automatic script. Other writings followed which apparently expressed in symbolic manner her secret wishes.

These findings led me to carry out some experiments in mental suggestion. The subject was told: "I am going to suggest to you mentally whatever comes into my head—just keep holding the pencil on the sheet of paper." Then, sitting opposite the percipient at a table covered with a tablecloth, I put a notebook on my knees, made my first pencil sketch, and began to suggest this mentally to her. I repeated this experiment four more times. The notebook, the pencil and my hands were securely concealed from the percipient by the table and the tablecloth. Control experiments carried out when the subject had left showed that it would not have been possible for her to recognise the figures drawn, either directly by sight, or by inference from the movements of the sender's shoulder: no one ever succeeded in doing so, whereas under just those conditions, the following results were obtained with C. as percipient:

Task 1. The figure 8 is transmitted. The percipient draws overlapping trapezia and triangles; the lower one is crossed by a line which forms a figure, similar to an 8. On the whole one must consider the task as having been carried out, but very indifferently (Fig. 9).

Task 2. Suggested: a circle with a small cross on the top (astronomical sign of the planet earth). The percipient drew an oval with two lines in it, crossing each other. The task may be considered carried out (Fig. 10).

Task 3. Suggested: the letter "y" (Fig. 10). Percipient accurately carries out task and writes a "y" (Fig. 10).

Task 4. The sender repeats the first assignment, which was not achieved— he again suggests the letter 8. The percipient writes a word which can be read as "one," then plainly the word "seven," then a crooked line and finally, distinctly, in the lower right hand corner, a capital "B." This is the end and there is no room for more.

The writings of the percipient could be interpreted as 1 + 7 = 8, but this would be an unjustified assumption. Nevertheless, the result of this experiment is more definite than task 1 (Fig. 11).

Task 5. It occurred to the sender to suggest the unusual name of a young girl whom he knew well, and who is now dead. Without saying anything to the percipient he writes in his notebook under the table "Elma." The percipient writes something that is apparently not what is suggested, namely "Mamara," then without interruption "um" with a downward stroke, then on a new line without dividing up the words "Sleeping I want," which probably meant "I want to go to sleep"—in fact at that time she did in fact look as though she would there and then go into hypnotic trance. I brought her back to normal by calling out sharply, and she absolutely refused to continue the experiment that evening. The task was not accomplished in a direct sense. It should be noted however that the word "Mamara" would be a slight disguise for the name "Tamara" (cf. the first writing in Fig. 8, "I want to live peacefully, I am fed up with ta . . .") Does not "ta" stand for "Tamara" and the unfinished word "um" for "will die"? [The Russian small hand-written *t* is in fact the same as the Roman written italic *m* as an alternative to a small T, and rather similar to a written m; "umret"—will die, "umeret"—to die]. (Fig. 12)

It should be added that C. did not know the dead Elma, but she did know a friend of ours—Tamara—whom she disliked. If one follows these surmises of mine (i.e. that *Mamara* means *Tamara*, and that *um* means *will die*) then the experiment which at first had seemed to be a failure might, nonetheless, be full of concealed meaning: (1) a woman's name (though a different one), (2) the idea of death, were both guessed. It would appear as if the thought of the death of an unknown girl, sensed telepathically, was subconsciously transposed into the thought of the death of a living and hated Tamara.

However, let us now return to the interrupted description of the experiments of the Commission for the Study of Mental Suggestion.

The experiments of the second series were conducted by one sender, one percipient and two assistants (one to help the sender, the other the percipient). Before the experiment the sender's assistant, without the knowledge of any of the other participants, chose three objects which bore no resemblance to one another as regards shape, material, colour, or purpose. The sender and the percipient went into separate rooms, each accompanied by his assistant, and the doors were closed. A piece of well illuminated white paper was placed on a table in front of the sender. His assistant placed on the paper the object which was to be fixated, and mentally suggested to the percipient. Each bout of suggestion lasted for about three minutes. The sender informed his assistant of the thoughts and images that came to him by way of association with the suggested object. The assistant timed the onset and termination of fixation and suggestion by means of a stop watch, and made notes of the position of the object and the verbal report of the sender. The percipient, meanwhile, remained with his eyes closed during the entire time of the experiment and, like the sender, informed his assistant of all the thoughts and images that occurred to him, and these were similarly noted down by his assistant who also used a stop watch. The watches of both assistants were synchronised to within an accuracy of 5 seconds. After the end of each experiment the notes were compared. The additional observations made by sender and percipient after the end of the actual experiment were then added to the record. These observations proved to be useful in the course of subsequent similar investigations as regards the interpretation of the experimental material.

The sender's part in the experiments conducted according to this method consisted in his visual concentration on the object of suggestion, steadily examining the whole object as well as its details, but excluding from his mind as far as possible any associations that the object might bring to mind, and indeed any extraneous thoughts whatever. Nevertheless, the visual imagery that came to the percipient under these experimental conditions usually had no complete or even appreciable similarity with the object of fixation. Frequently the percipient "perceived" only separate parts, details or symbols of the object, which could not be said to characterise the object as a whole. For example, an object such as the uncovered dial of a pocket watch would call up in the percipient a partial image of this object—a circle, a metal ring, a thought about figures, the image of an open compass, the hands of the watch, a working mechanism etc. Glass images usually elicited images of the surface of water or ice.

Thus, in one experiment (18/1/1923), the sender's object of suggestion was a strongly illuminated block of cut glass in the shape of a pyramidal polyhedron. The percipient saw "reflections in water—sugar loaf—snowy summit—iceberg, icefloes in the north illuminated by the sun—rays are broken up."

As a result of a careful examination of the percipient's verbal report, the members of the Commission reached the conclusion that "the perception of mentally transmitted images always enters the subconscious of the percipient in the form of hidden [unobserved—L.L.V.] processes; a secondary reflected process appears in his consciousness." Consequently, the image actually perceived is usually distorted to a greater or lesser extent.

The members of the Commission noted that unobserved subconscious experiences play a great part, not only in the process of perception (of images), but also in the sending process. This conclusion they formulated as follows: "That to which the voluntary and deliberate attention of the sender was not directed was more easily transmitted to the percipient: this would somehow or another attract the undirected (reflected) attention of the percipient and subsequently elicit in him a more or less pronounced conscious reaction—a spontaneous thought, a detail of the fixated object unexpectedly noticed, a desire to conceal something, etc." This is exactly why "more frequently it was not the actual image suggested that was transmitted but images that have some fortuitous association with it. This explains a fact frequently noted by other experimenters and confirmed by us, that successful transmission very often does not depend on the sender's will: one can gaze at the detail of an object, even a picturesque and prominent detail, as long and as attentively and concentratedly as one likes, full of eager desire to transmit it to the percipient, and this may produce no effect whatever. Conversely it frequently happens that details and associated images are transmitted to which the sender paid no attention while fixating the object, and which were only noticed by him after the experiment when he learned of the percipient's reactions by reading the experimental report."

This conclusion is supported in the report of the Commission by the following experiments:—[6]

Experiment 1. Object of fixation: a small oval portrait of a woman in a glazed, dark red, square leather frame. During fixation the sender noticed on the glass the reflection of the filaments of an electric light bulb, reminiscent of the contours of the Roman letter "N," and he said to his assistant: "Napoleon—the letter 'N' flashed by." Within 25 seconds of this remark of the sender's, the percipient said (and this was recorded by his assistant): "A palm tree, a wreath, the word 'consul'"; after another 25 seconds: "I see either Napoleon or Vespasian." (Experiment 18/1/1923)

Experiment 2. Object of fixation: a white folding bone knife (for cleaning nails). Almost immediately after the beginning of "transmission" the percipient said: "The figure of a lion made out of ivory—as a paper knife the handle of which—a lion's head." The sender said that the object of fixation reminded him of an object he possessed—a round powder box made out of ivory, on the lid of which there is a plain picture of a lion's head. (Experiment 22/12/922)

In these experiments perception followed transmission almost instantaneously. Such cases are called "synchronous (simultaneous) intercerebral (brain-to-brain) transmission and perception" by the members of the Commission in their report. However, cases of asynchronous (delayed, not simultaneous) perceptions were likewise observed. Such a delay in eliciting the image seemed a chance happening with some percipients, but with others such a delay was a permanent feature.

It was possible to cause such a delay artificially by the following experimental procedure: It was suggested to the percipient, who was in a hypnotic trance, that he would only perceive the images starting from a certain moment. In the experiments of the Commission this moment was signalled to the percipient by placing a wax helmet on his head; any other mild stimulus could act in the same way. And, sure enough, perception thereupon ceased and remained latent until the specified moment, just as though the transmission was received, not from the sender but from the helmet on the head, a significant feature in common with the formation of delayed conditioned reflexes. The third series of experiments was devoted to the study of such asynchronous transmissions and perceptions.

During the two years of the Commission's work 269 experiments of all three series mentioned above were conducted; about 900 different objects and pictures were used for transmitting by mental suggestion. The numerical results (in percentages) of all the experiments are as follows:

Table 2

Series	No. of Experiments	NO. OF EXPERIMENTS (PERCENT) OF WHICH		
		Perception More or Less Accurate	Perception Not Complete or Symbolic	No Perception or Doubtful
1	68	20	22	58
2	118	14	40	46
3	83	11	27	62

Results of experiments in mental suggestion of visual images.

It should be taken into account that the choice of objects for mental suggestion was not in any way limited in the above experiments. In other types of experiments the choice is limited involving, for example, a choice of 52 in the case of playing cards, 30 in the case of the letters of the Russian alphabet, 10 if numbers are used and 5 for Zener cards.[7] In our experiments the probability of chance coincidence is very low, and therefore the percentages shown in the table for complete and incomplete perception must be considered very high.

The actual experimental set-up in which two assistants noted down the immediate and subsequent verbal reactions of sender and recipient respectively is of interest. An analysis of the results thus obtained enabled the members of the Commission to determine the unobserved (non-deliberate, subconscious) psychical elements in the process of transmission and perception. The members of the Commission were pleased to learn that foreign observers, such as E. Osty, R. Warcollier,[8] Desoille[9] had reached the same conclusions as regards the role of subconscious factors in experiments on mental suggestion. Desoille expressed this with especial clarity, and elaborated it by means of a drawing (Fig. 13). I here quote an extract from the work of this author: "There exist in all 4 psychological possibilities for effecting thought transmission, i.e. for the realisation of an act of will which the percipient P perceives as transmitted by the agent A. These 4 ways are the following:—

1. Direct transmission from the conscious mind of A to the conscious mind of P.

2. Transmission from the conscious mind of A, first to the unconscious of P, and thence to P's consciousness.

3. Transmission from the conscious mind of A to his own subconscious mind and then to P's conscious mind.

4. Transmission from the consciousness of A, first to his own subconscious, then to the subconscious of P, and finally to P's consciousness."

Desoille observes that it is not at present possible to decide which of these interconnections is in fact the one that occurs in reality, but there is a good deal to be said in favour of the fourth alternative. My own view is that all four possibilities are in fact actualised; the problem is to decide in any given instance which is the correct one. The first and the last represent extremes, the second and third intermediary stages.

This seems a good place to mention that the concepts of "conscious," "subconscious" and "unconscious"[10] which I have employed for the categorisation of mental phenomena are frequently connected with theories of the psychoanalytic school of Sigmund Freud which are unacceptable to Soviet readers. In point of fact Freud, in his numerous works, enormously exaggerates the role of the unconscious and attributes mystical properties to it. But it would be wrong to assume that subconscious mental phenomena do not exist—that they exist only in the imagination of Freud and his followers. I give below what has been written on this point in a recently published book devoted to a severe criticism of the Freudian position from a dialectical materialist point of view: "Freud is right as regards one thing: the unconscious exists. It lives, acts, affects conscious processes, and that not in the form of some 'purely physiological' temporary nervous connections which are blocked for a time. It is a living realm full of sense the effects of which are felt by everyone of us. That which we know but do not at the moment remember has a definite meaning content which, in principle, does not result in the stimulation or blocking of cortical cells."[11]

Psychologists and physiologists of the Pavlov school (at least some of them) do not deny the existence of the subconscious realm. They write: "In going through the voluminous literature we find clinical material of the utmost interest for psychiatry, neuropathology and psychotherapy, where literally every page provides one with valuable empirical material, with striking cases of descriptions of various types of disorders, from which there emerge most plainly the facts of subconscious phenomena."[12]

The initial source of such phenomena may be not only unfelt interoceptive stimulation, but also sub-threshold exteroceptor stimulations of the external sense organs. The subconscious is fed by consciously perceived experiences which are subsequently, for some reason or another, withdrawn from consciousness, "pushed out" into the sphere of the hidden (latent) memory. In the article cited the authors, who are physiologists, show convincingly that the physiological basis of subconscious mental phenomena is in accord with the laws of higher nervous activity laid down by I. P. Pavlov and his school.

The possibility that the course of mental suggestion experiments may be interfered with by subconscious mental factors in the hidden memories, both of sender and of percipient, requires special vigilance on the part of the experimenter. At times an experiment that appeared completely successful proves, after sender and percipient have both been questioned, to be only of doubtful validity. The converse also can happen: an experiment which seems unsuccessful assumes considerable interest after both sender and percipient have been questioned. I will here cite a few examples of this type.[13]

On December 24, 1932 between 11 and 12 p.m. I carried out the following experiments. Sender: the author; percipient: his wife, T. B. The percipient, while lying on her back, is given the instructions to make herself quite passive and say whatever comes into her mind. The sender, sitting behind her head, has on his knees a notebook and draws in it the visual pictures which are being mentally suggested.

Experiment 1. The sender draws a pair of scissors. The percipient's verbal response: "I am plunging into something—a uniform with buttons and a cap—a bow tie." Question by the sender: "What is this tie you are talking about?" Answer: "Your tie" (tied in a bow "butterfly" type). Percipient, immediately: "And now, about scissors." Question: "How did the word 'scissors' come into your mind?" Answer: "I found myself mentally pronouncing the word 'scissors'; it just came to me."

When a certain time had elapsed after the experiment, the percipient recalled that she had recently read a book on telepathy in which, amongst a number of other experiments, there was a description of a successful instance of thought transference, the object of which was an open pair of scissors. This, as well as the sender's tie which resembled an open pair of scissors, could have suggested the correct solution to the percipient.

Experiment 2. Experimental conditions: as above. Instruction: "I am thinking of a two figure number. Tell me even if you see a one figure number." (I thought of the number 25.) Within 30 to 60 seconds the percipient says: "the number 25 keeps recurring over and over again." To the question "How did you come to think of the number 25?" she replied: "The thought of 25 appeared immediately after the beginning of the experiment, but it seemed inconvenient to say it right away. It is my favourite number." (The sender either did not know this or had forgotten it.) It should also be added that this was the 24th of December, the eve of the 25th.

These circumstances reduce the significance of the experimental results, so convincing at first, to nil. The sender could subconsciously have selected the number 25 because tomorrow was December 25th. The percipient could have picked on this number either for the same reason or because it was her "favourite number."

The results of experiments on the mental suggestion of visual images invariably confront the experimenters with riddles of this sort. The evaluation of the results of such experiments is in many cases, ambiguous: what seems convincing to one person seems totally unimpressive to another.

Hence the experiments referred to in this Chapter are of little use in proving the factual occurrence of mental suggestion: they will not satisfy the sceptics; but for researchers who do admit the existence of telepathic phenomena, experiments in mental suggestion of visual images can, better than any others, throw light on the optimal conditions for telepathic transmission and perception, as well as on psychological mechanisms of great depth involved in mental suggestion. This is exactly why the above methods (transmission of scripts) are so often used. However, the results thus obtained are too complex, they only admit qualitative evaluation, and consequently are only of slight use for the purpose of statistical analysis, or for carrying out those types of psychophysical research which are of the greatest interest for us. And this is precisely why, at the suggestion of Academician V. F. Mitkevich, we adopted a method of mental suggestion using more simple visual experiences—the experiences of white and black.[14]

Mitkevich's suggestion was usually carried out by us in the following rather simple way: the sender was in the same room as the percipient, a few metres behind his back. The sender had at his disposal a box containing six small balls, three white and three black ones. After shaking the box he took out one of the balls at random. The sender then tried to visualise with as much clarity as be could the "colour" of the ball in question, white or black, and to transmit this image to the percipient. The ball was then replaced in the box.

After a signal of "ready" from the sender, the percipient would give, usually within a few seconds, a verbal response of "white" or "black." The experimental data (the "colour" transmitted and the percipient's guess) were immediately entered in the experimental record. In every experiment with a given percipient 40 mental suggestions were "sent" in succession with a short interval after every 10 calls. After a rest period of 10 to 15 minutes the experiment was resumed and another 40 calls were made. There were thus 80 tests in all, and from that number the percentage of correct answers was worked out.

In some of the experiments what was transmitted to the percipient was not the concept "black" or "white," but rather the personal feeling tones associated with them. In such cases a simple little apparatus, made for the purpose by V. F. Mitkevich, was placed in front of the sender; this consisted of two discs on a rotating vertical rod, one white and one black. The sender himself sets the apparatus in motion; when it stops rotating he finds himself confronted with a black or a white disc which he fixates, and whose "colour" he tries to transmit to the percipient (Fig. 14).

By the rules of probability 40–50 such experiments are adequate to get 50 percent correct answers which depend on chance alone. If the experiments, however, yield a percentage of correct answers higher than 50 percent, it is then possible to speak of a greater or lesser probability of the incidence of mental suggestion in these experiments.

The two great points in favour of this method are that each individual test takes little time, and that results can be assessed in terms of simple alternatives: right or wrong, yes or no, instead of a qualitative evaluation which applies to the majority of other methods of studying thought transmission. Consequently it is possible, by means of this method, to collect quickly a large amount of numerical material which is suitable for statistical analysis. However, this method also has its drawbacks: senders experience difficulty in visualising black and white with sufficient intensity, and this difficulty is aggravated by the necessity of having to switch many times from "black" to "white," and vice versa.

Nevertheless, B. F. Mitkevich claims that in his experiments this method usually yielded positive results, i.e. more than 50 percent correct answers. We decided to verify this contention for ourselves, using the above described method on a sufficiently large number of experimental subjects, and subsequently trying to assess its possible usefulness in solving the problem of the psychophysical experiments we had undertaken.

To this end my collaborators and I carried out 26 experiments (80 calls each) on 22 subjects, some in good health, others suffering from various hysterical and neurotic complaints. The results obtained are to be found in Table 3.

Table 3

Serial No.	Subject	No. of Correct Replies	Percentage of Correct Replies
1	D.G.P.	60	75.00
2	D.P.V.	40	50.00
3	P.A.P.	51	63.75
4	P.A.A.	64	80.00
5	T.I.F.	40	50.00
6	I.E.M.	41	51.25
7	T.S.D.	35	43.75
8	S.M.A.	40	50.00
9	I.E.M.	40	50.00
10	B.R.M.	44	55.00
11	D.V.E.	32	40.00
12	I.E.M.	37	46.25
13	K.S.V.	42	52.50
14	N.A.I.	45	56.25
15	L.Y.I.	42	52.50
16	I.E.M.	35	43.75
17	F.K.G.	62	77.50
18	V.K.I.	41	51.25
19	B.E.M.	47	58.75
20	L.A.F.	37	46.25
21	B.M.M.	53	66.25
22	V.B.A.	42	52.50
23	K.N.P.	52	65.00
24	I.E.M.	45	56.25
25	B.G.P.	40	50.00
26	P.S.A.	49	61.25

Number of subjects: 22, M = 44.46, percentage mean = 55.58

NOTE: I.E.M. is our experimental subject Ivanova, F.K.G. is Fedorova. 5 experiments were conducted with I.E.M. (Nos. 6, 9, 12, 16, 24). All other subjects: 1 experiment each.

Results of the experiments in mental suggestion of white-black by the Mitkevich method.

An examination of the numerical data given above will show that the highest number of correct answers was 64, and the lowest 32. The arithmetic mean of the scores is thus 44.46 instead of 40 (55.40 percent instead of 50 percent) which would have been expected from probability calculations. Thus only a trifling increase over the number expected by chance alone was obtained—an increase of 4.46 (5.60 percent).

It should be pointed out that such a result was not unexpected, so far as we were concerned: such meagre results are the rule in experiments carried out without prior selection of subjects. Richet's experiments with playing cards may serve as an example. In these experiments the percipient had to pick out a certain card which was in the sender's mind out of a number of cards scattered over the table. In all, 2,997 experiments were carried out and 789 correct answers were obtained, whereas the number calculated from probability theory was 732. Thus, even in Richet's experiments there was only a trifling advance over the number of correct guesses expected by chance alone.[15]

The question that has to be answered is: can the 4.46 difference which we obtained be regarded as unquestionably exceeding the limits of chance errors which are theoretically possible under our given experimental conditions? In order to answer this question we took advantage of the most simple statistical technique. For our average arithmetic mean ($M = 44.46$) we compute the average error, m, which appeared to be $m = 1.16$. Applying the formula $M \pm 3m$ we see that the value M, depending on a single chance, may vary to the extent of $3m = \pm 3.48$, i.e. from 41.00 to 47.96.

In other words, even on the worst of the probable expectations (41.00), the result obtained in our experiment still somewhat exceeds the average expected on probability theory (40), and this means that the excess of the results obtained by us over those computed as expected theoretically if chance alone were operative, although small, is statistically significant.

It follows that (if one accepts that "prompting" cues have been completely excluded in our experiments) mental suggestion has been statistically justified as being a factor enabling the percipient to guess the visual targets, black and white. The value of the method devised by Academician Mitkevich can be supported by a perusal of the results obtained in some other experiments. By way of illustration I give here the following two examples:—

Sitting 1. 1/12/1933. Subject L.V.S. (24 years old, had previously suffered from hysterical disorder, easily hypnotised) is put to sleep (somnambulistic type of hypnosis) and in a chamber (No. 2) equipped with a wooden cover (see below); the sender (A. V. Dubrovsky) is a few metres from the chamber; also present are L. L. Vasiliev and S. F. Tomashevsky.

The sender has the idea of transmitting during the whole of the sitting (40 experiments) nothing but "white."

In the first 10 experiments the percipient gives only 4 correct answers; in the second he gives 5; in the third 10 he gives 8 and finally in the fourth 10 he gives all 10. This would appear to indicate that a mental suggestion of "white" repeated over and over again during the sitting gradually creates in the percipient a stable orientation on "white." Sure enough, when in the experiment that followed the experimenters replaced "white" by "black" the subject, as though by inertia, repeated "white" three times, and then switched over to "black," naming it several times in succession. It should be added that in the usual set-up of this experiment in which "white" and "black" were mixed at random, this subject, like other percipients, never gave the same colour more than 3 times in succession.

Sitting 2. 23/4/1934. Subject K. G. F. (35 years old, hysteric) is in the Faraday Chamber (No. 1) in a hypnotic state. Dr. Dubrovsky, the observer, is with her; the door is open. The sender, I. F. Tomashevsky, as well as others present—N. I. S . . . ov, R. I. Skariatin and L. L. Vasiliev—are outside the line of sight of the subject. The suggestion "white" or "black" is effected by means of the Mitkevich rotating discs. The sitting is held to demonstrate this method to N. I. S . . . ov who was present at our experiments for the first time. The experiment is conducted, and the notes are made, by S . . . ov himself.

Out of 10 tests the subject gave 10 correct responses in succession. This is a somewhat unusual occurrence in control tests of chance expectation, the probability being calculated to be 1 in 1024.[16]

All these preliminary tests, as well as others not quoted here for lack of space, suggested to us that the Mitkevich method might be suitable for psychophysical experiments. Such experiments were subsequently carried out by us under the following conditions:—

The subject, who remained awake, was put into a metal screening chamber (No. 2), the sender was at a distance of 2 to 3 metres from the chamber. Under screening conditions 10 tests were carried out, "black" and "white" being suggested. Subsequently the upper part of the chamber was raised by means of the pulley and was replaced by a wooden cover. Under these conditions, i.e. without screening, another 10 tests were carried out. The third set of 10 was again conducted under conditions of metal screening, the fourth without screening once more. The subject then had a 10 to 15 minutes rest, after which the second part of the sitting was conducted in precisely the same sequence as the first. Consequently, 40 answers of the subject were given *with* screening by iron, and 40 *without*. The number of correct answers obtained with screening as well as those obtained without screening were subjected to statistical computation.

In all, 22 such sittings were held with 13 subjects (healthy as well as sick). The total résumé of the data is given in Table 4.

The summary shows that experiments without and with screening gave an average slightly exceeding the number 20 (the chance expectation). In the former case (without screening) this excess over chance expectation is significant ($M_1 = 22.45 \pm 0.88$), and in the latter (with screening) it cannot be considered significant ($M_2 = 21.50 \pm 0.99$). Moreover, M_1 is noticeably higher than M_2.

So the factor of mental suggestion, which apparently played a part in our experiments without screening, was not present in experiments with screening.

If this is so, we can draw a most important conclusion, namely that screening by means of metal impeded the transmission of mental suggestion from the sender's brain to that of the percipient.

But before establishing such a conclusion, it is first necessary to determine whether the difference between M_1 and M_2 is significant. This is not difficult, and may be determined by calculating the so-called coefficient of significance of the difference (t) by the formula often applied in statistics:—

$$t = M_1 - M_2 / \sqrt{(m_1^2 + m_2^2)} > 2$$

Table 4

Sitting No.	Percipient	NUMBER OF CORRECT REPLIES (POSITIVE RESULTS) Without Screening (Out of 40)	With Screening (Out of 40)
1	I.E.M.	23	25
2	B.T.E.	25	18
3	I.E.M.	25	24
4	B.T.E.	25	24
5	B.T.K.	26	24
6	I.E.M.	18	27
7	M.R.I.	25	21
8	I.E.M.	23	26
9	S.V.M.	20	12
10	I.O.V.	17	23
11	F.K.G.	28	23
12	I.E.M.	19	21
13	V.E.P.	18	18
14	K.G.Z.	23	16
15	K.G.Z.	20	19
16	U.E.P.	20	23
17	K.G.Z.	16	11
18	E.T.V.	25	19
19	S.L.V.	34	23
20	S.L.V.	20	26
21	E.T.V.	23	20
22	S.L.V.	21	30

Number of subjects: 13.

$M_1 = 22.45 \pm 0.88$ $\quad$ $M_2 = 21.50 \pm 0.99$

$\sigma = 4.12. \pm 0.62$ $\quad$ $\sigma = 4.63 \pm 0.70$

NOTE: I.E.M.: Ivanova, F.K.G.: Fedorova. 5 experiments were carried out with subject I.E.M., 3 with K.G.Z. and S.L.V., and 2 with B.T.E.

Results of experiments of mental suggestion carried out by the Mitkevich method, percipients being asked to guess "black or white?"

As is well known this formula is interpreted as follows: the difference between 2 arithmetic means, M_1 and M_2 may be considered significant only if the difference between them divided by the square root of the combined squares of average errors (m_1 and m_2) is greater than, or at least equal to, 2. If, however, the above mentioned ratio is less than 2, then the difference obtained between the results of the two series of experiments is not significant and no use can be made of it.

By putting into the formula the values M_1, m_1, and M_2, m_2 obtained in the experiment we have

$$t = 22.45 - 21.50 / \sqrt{(0.88^2 + 0.99^2)} = 0.72 << 2*$$

As we see, the coefficient of the significance of the difference between M_1 and M_2 is considerably less than 2. Consequently the difference of results obtained by us in experiments with screening and experiments without screening cannot be considered as significant, and one must presume that it depends on chance. In view of the rather small number of results (the arithmetic mean and average errors are computed from 22 experiments) we employed the improved statistical method of Fisher.[17]

The calculation of the biometrical constant was made by the non-weighted method. The mean square deviation was computed by the usual formula, and was then multiplied by $\sqrt{N} / (N - 1)$ in order to compensate for the small number of results (N being the number of experiments). Errors in all cases are mean errors. The assessment of the significance of the difference, t, was made by means of the Student t-test. The conclusion was deduced on the basis of the probability of the null hypnothesis, designated as P. This value is obtained from the above mentioned tables (see Fisher), entered with the values of t and the number of degrees of freedom ($n_1 + n_2 - 2$).

Comparative M:

$$t = 22.45 - 21.50 / \sqrt{(0.88^2 + 0.99^2)} = 0.72*$$
$$n_1 + n_2 - 2 = 42; 0.4 < P < 0.5$$

Comparative σ:

$$t = 4.63 - 4.12 / (0.62^2 + 0.70^2) = 0.5 < P < 0.6$$

A comparison of the averages of the absolute numbers gives a probability on the null hypothesis: $0.4 < P < 0.5$; for mean squares $0.5 < P < 0.6$. It follows therefore that the effectiveness of the screening was not established, since

---

*See Introduction, page xli.

the difference between the means of the two series of experiments is a chance one, and the probability of the null hypothesis exceeds the conditions of significance, now accepted in statistics, by many times: $P < 0.05$.

Thus, despite our first impression, we reach the statistically well grounded conclusion that screening in this case did not affect the transmission of mental suggestion; in other words, the metal screen did not prevent the passage of the energy which mediates mental suggestion.

It would, however, be wrong to consider such a contention as proved by the experiments just described. It would be a mistake if only because the part played by telepathic influence in this instance is very trifling, and indeed for the experiments with screening it is not significant.

In order to substantiate our preliminary conclusion we were thus forced to employ another method of mental suggestion—one which yields far more significantly positive results: this turned out to be the procedure of putting to sleep, and awakening, by means of mental suggestion.

6

Mental Suggestion of Sleeping and Awakening

The first experiments of inducing sleep and waking up by means of mental suggestion were carried out in Le Havre by the celebrated psychiatrist Prof. Pierre Janet and his colleague, Dr. M. Gilbert in 1886. At distances ranging from a quarter up to one mile it was found possible, by means of mental suggestion, to induce in the percipient, Léonie B. (a healthy peasant woman 50 years of age), a condition of hypnotic sleep at any time selected at will by the experimenter who acted as sender. Out of 25 experiments 19 were wholly successful, the remainder being doubtful resulting only in drowsiness or had no effect.[1] These findings were vouched for by an authoritative commission and were repeated in Paris by Charles Richet, with the same subject. The sender, (Richet), tried to put the subject to sleep by means of mental suggestion, while she was separated from him by a distance of 1 to 2 km. The experiment was successful in 16 cases out of 36: the subject fell asleep soon after the onset of suggestion (for details see Appendix D). In more recent times Dr. S. Alrutz whom we have already mentioned, reported similar experiments, in his memorandum submitted to the First International Congress of Psychical Research in 1921:[2] "I can very easily, by means of telepathy, put my subjects into a hypnotic sleep, and find it particularly easy to wake them up again in the same way" (p. 282).

At the Second All Russian Congress of Psychoneurology at Petrograd in 1924 Prof. K. I. Platonov (see Appendix E) reported that he had noticed that some of his patients could be put to sleep and woken up again by having the eyes of the experimenter focussed on them—if this was accompanied by a mental instruction to go to sleep, or to wake up: subjects could thus be awakened either from normal or from hypnotic sleep. The lecture was accompanied by a demonstration there and then of these phenomena, using one of his patients as a subject. The experiment succeeded even when Prof. Platonov was outside the lecture room, that is when the subject was out of his field of vision. He said that in only 10 or 12 subjects out of 300 with whom he had experimented in this connection was he able to demonstrate the phenomenon with any degree of regularity.

Prof. Platonov's experiments, using the same experimental subject, were repeated by V. M. Bekhterev jointly with the present author. In these control experiments it was noticed that this subject, Mikhailova, once the suggestion to go to sleep was given, soon fell into a condition of auto-hypnosis; that is, a conditioned hypnogenic reflex was established similar to that described by Dr. B. N. Birman (a collaborator of I. P. Pavlov's) on dogs.[3]

I. F. Tomashevsky, who took part in our joint researches on mental suggestion at the Leningrad Institute for Brain Research, claimed that he himself had carried out similar experiments on the subject Krot, 21 years old, as far back as the '20s: with this subject it had been possible to induce hypnotic sleep at the most inopportune times for sleeping, for instance while she was walking, talking, etc. It is also interesting to note that, prior to the mental suggestion experiments, the subject had never been subjected to the usual verbal suggestions resulting in hypnosis.

At the beginning of our joint researches in 1933 we were lucky enough to find two or three subjects capable of exhibiting the phenomena of going to sleep and waking up, in response to mental suggestion, in a sufficiently clear-cut form.

The experience thus acquired by us and by other researchers led us to subject this "hypnogenic" method of mental suggestion to a detailed study for the purpose of testing whether or not it might be useful in the design of our projected psychophysical experiments.

Ivanova and Fedorova, both hysterical patients whom we have already mentioned in Chapters 3 and 4, proved to be suitable for our purpose and the experiments carried out with them provided our basic empirical material. A small number of experiments were also carried out with a third subject E. S. (normal, 27 years old). In these experiments our scientific collaborators, I. F. Tornashevsky and Dr. A. V. Dubrovsky, acted as senders.

In the first version of these experiments[4] the procedure was as follows: the subject was either sitting in an arm-chair or lying on a bed; there was one observer with her, and another collaborator timed the subject's reactions by means of a stop watch. The sender (who was out of sight of both subject and observers, either in the same room or in another room) began to exert mental suggestion, attempting to induce sleep at a moment unknown to either subject or observer, the watches of sender and observer having been compared beforehand. During suggestion the sender tried to reproduce with the greatest possible vividness feelings usually experienced when falling asleep, and to associate these feelings with the image of the percipient while mentally conveying the command "go to sleep!"

If this suggestion was successful, the sender, after a certain time (again at a moment unknown either to observer or subject) began to give suggestions of awakening, proceeding in the same manner as when putting to sleep. In the course of a single experiment of this type the same procedure of awakening and putting to sleep was often repeated several times with different intervals, ranging from a few minutes to an hour or more.

In order to make this "hypnogenic" method more rigorous we introduced various pieces of technical equipment. Thus, a microphone was placed in the same room as the subject, connected by an amplifying system to a loudspeaker in another room, some distance away. This enabled the experimenters, who were in this other room, to listen in on the subject's "audible behavior" (her utterances in hypnosis, her sighs, her conversations with the observer if the latter was with her) without her knowledge. But special importance must be attached to the pneumatic registration device of the subject's rhythmic movements which we introduced. For this purpose we made use of a somewhat altered version of the Marey apparatus which is so often employed by physiologists. Essentially our method was as follows:

A rubber balloon filled with air was placed in the subject's right hand and tied across the back of the hand for greater security. By means of a rubber tube which ended in a metal tube this balloon was connected with a Marey capsule, generally placed in another room. The Marey capsule consists of a cylinder covered at the top by a movable rubber membrane which rests on a light lever. The lever of the capsule with its free end touches the soot covered ribbon of the rotating drum of a kymograph. It will easily be seen that, with this set-up the slightest pressure exerted by the subject on the balloon is pneumatically transmitted to the Marey capsule, thus causing a bulging of the membrane and the raising of the registration lever which will trace a record on the cylinder of the kymograph in another room corresponding to the hand movements of the subject.

The experimental procedure was as follows: the subject, who was awake, was instructed to press the balloon rhythmically. This the subject was apparently able to do without effort or fatigue for a considerable time. At a moment in the course of the experiment unknown to the subject, the sender would start his mental suggestions to the subject to go to sleep; at that point he, or his assistant, switched on the electromagnetic recorder which registered on the same kymograph the moment of the onset of the suggestion to go to sleep.[5] While the percipient remained awake the pneumatic registration of her movements continued, but as soon as the suggestions took effect, and the percipient fell into hypnotic sleep, the movements ceased at once and, so long as sleep continued, the registration apparatus recorded an even line.

As soon as the sender was about to begin the suggestion of mental awakening he again switched on the recording apparatus. When this suggestion took effect, i.e. when the percipient woke up, she immediately and without special instructions resumed the compression of the balloon which had been interrupted during hypnosis. Generally, as a result of post-hypnotic amnesia, the subject did not notice that during the hypnotic state she had not continued to press the balloon.

By this method we succeeded in making (in our experiments with the three subjects mentioned) a large number of kymographic records which, so it seems to us, completely establish objectively the phenomena of mentally induced sleep and awakening. By way of illustration I am giving the kymographic record obtained in an experiment, on 19/12/1933, with Fedorova as percipient and Tomashevsky as sender (Fig. 15).

An examination of this kymogram shows that, at the beginning of the recording, the subject was awake. At a certain moment in the experiment, indicated by the lowering of the line of the lever (first 3) the sender mentally begins to send the subject to sleep. The record shows that this suggestion took effect almost instantly: the swings of the curve (i.e. compression of balloon by the subject) stop (first at Γ: hypnosis). After a certain time the sender transmits the mental order "wake up!" (first at Π on the line of the register), and the swings of the curve are resumed soon afterwards: the subject wakes. After a certain time the sender again starts mentally suggesting sleep (second 3 on the line of the register) and the subject instantly falls asleep. The sender starts waking her up again, and the swings of the curve appear once more.

In this experiment it was possible, in a short period of time, to obtain three inductions of sleep and three wakings-up which occurred within seconds (first three registrations) and within tens of seconds (second lot of three registrations) after the corresponding mental suggestions. It is hardly possible to explain these results as being due to fits of spontaneous hypnosis (so-called auto-hypnosis) and awakening accidentally coinciding in time with the corresponding mental suggestions: the probability of such a coincidence is too small. This supposition must also be rejected because nearly similar results were obtained by us in dozens of instances in numerous experiments on three different subjects.

Fig. 16 is a photographic reproduction of a whole kymographic tape on which are entered the results of three experiments with our three subjects (22–23 Feb. 1934); in these experiments percipient and sender were in different rooms. Let us explain some features of this kymograph.

Experiment 1. Subject: Fedorova. Sender: Tomashevsky. The subject starts compressing the balloon at 9:55; at 10:12 the sender begins the suggestion "go to sleep"; the subject goes to sleep within 6 minutes, at 10:18. At 10:50 the same sender begins the mental suggestion "wake up" and within as little as 30 seconds (10:51, i.e. 30 seconds) registration is resumed—the subject comes out of hypnosis.

Experiment 2. Subject: Ivanova. Sender: Dubrovsky. At 11:35 the sender starts mental suggestion to go to sleep which takes effect within as little as half a minute (the compression of the balloon ceases). At 11:48 a mental order is given to wake up and this order is obeyed at 11:49 (the swing of the curve is resumed).

Experiment 3. Subject: E. S. Sender: Tomashevsky. Mental suggestion to go to sleep begins at 11:26; within 2 1/2 minutes hypnotic sleep is induced. At 11:40 the sender starts mental suggestion to wake her up. There is a delay of 7 1/2 minutes before the subject wakes up. At 11:54 the sender gives a second mental order to go to sleep which is quickly complied with. At 12:04 the subject wakes up spontaneously without having been given any mental order to do so.

This is only one of many cases which could be cited. In some experiments we tried to replace the compression of the balloon by a registration of the breathing movements by means of the usual belt pneumograph. However, this method of registration proved to be less sensitive: the pneumogram altered but little at the onset and termination of hypnotic sleep, and was consequently a poor indicator.

A more suitable method proved to be the technique of registering the strength of an electric current through the hand of the subject (by Veragut's method) by means of a sensitive mirror-galvanometer with a simple mechanical device attached to it. In addition to the compression of the balloon with the left hand, non-polarisable electrodes, connected to an electric battery, were placed on the surface of the back of the subject's right hand. This was done in order to obtain a simultaneous registration of the galvanogram on the ribbon of the kymograph.

Fig. 17 shows the registration obtained in an experiment on subject A. A. when she came to the laboratory for the first time and demonstrated a susceptibility to being put to sleep by means of mental suggestion. The experiment was started at 8:45, mental suggestion to go to sleep was made at 9:10, and as early as 9 h. 11 min. 12 sec. the swings of the registering lever stopped: sleep had set in. The galvanogram altered at the same instant: it went up temporarily, and (and this is the most characteristic) it lost during the entire period of sleep its jagged appearance, and continued as an even line. From the point of view of method, registration of one of the vegetative reactions has considerable advantages. Responses such as the alteration of the galvanic currents of the skin are not under conscious control and cannot be counterfeited by experimental subjects. The method of measuring galvanic skin response in connection with experiments of mental suggestion was first applied in 1922 by Brugmans in his joint work already cited.[6]

However many separate examples of the empirical material selected by us we might cite, they would hardly seem sufficiently convincing: it is the law of large numbers that plays a decisive part in experiments of this sort. This is the reason why we consider it necessary to submit in figures all our experimental material.

From 1933 to 1934, 260 experiments in mentally inducing sleep and awakening on the subjects Ivanova, Fedorova and E. S. were carried out; of these 194 were accompanied by kymographic registration, of the rest notes only were made in the usual way. Out of the total number of experiments (260) the mental induction of sleep failed in 6 experiments, and mental awakening failed in 21. This amounts to 10.4 percent.

In order to make use of the most valuable experimental material, namely that which was kymographically recorded, in the most systematic manner, let us divide it into four series of experiments:

1. Experiments without screening, sender and subject being in the same room.

2. Experiments without screening, sender and subject being in different rooms.

3. Experiments in which the sender is placed in a perfect screening chamber, the subject being in the same or another room.

4. Experiments in which the sender is placed in a perfect screening chamber and the subject in a Faraday chamber (double screening).

Table 5
First Group of Experiments

Subject: Ivanova

Date	Serial No.	S—H Min.	S—H Sec.	W—V Min.	W—V Sec.	Sender
13/12	1,2	6	00	8	00	D
15/12	3,4	10	00	7	00	D
17/12	5,6	19	00	5	00	D
19/12	7	-	-	1	00	T
29/12	8	2	00	-	-	D
15/1	9	-	-	0	30	T
11/3	10,11	2	00	0	30	D
	12	2	00	-	-	T
14/3	13,14	5	00	0	30	D
21/3	15	-	-	1	00	T
2/4	16	4	00	-	-	D

$M = 6.25 \pm 2.06$ min. 2.94 ± 1.13 min.
$\sigma = 5.83 \pm 1.46$ 3.20 ± 0.80

Date	Serial No.	Subject: Fedorova S—H		W—V		Sender
		Min.	Sec.	Min.	Sec.	
16/12	1,2	1	00	1	00	T
	3,4	1	00	1	00	T
19/12	5,6	3	00	1	30	T
	7,8	0	30	0	30	T
	9,10	0	15	0	15	T
	11,12	0	20	0	20	T
16/1	13	-	-	2	00	T
21/1	14,15	6	00	6	00	T
	16,17	1	00	6	00	T
25/1	18	-	-	2	00	D
21/2	19,20	6	00	1	30	T
8/3	21,22	3	00	2	00	T
13/3	23,24	3	00	1	00	T
	25,26	1	00	0	30	T
17/3	27,28	2	00	1	00	T
28/3	29	-	-	1	00	T
1/4	30	5	00	-	-	T
11/4	31,32	1	00	0	30	T
	33	0	00	-	-	T
14/4	34	-	-	4	00	T
22/4	35	0	30	-	-	T
26/4	36,37	1	00	0	18	T
	38,39	1	00	0	50	T
30/4	40	17	00	-	-	T
	41	16	00	-	-	T
5/5	42,43	2	00	3	00	T
	44,45	2	00	1	00	T
	46	2	00	-	-	T
9/5	47,48	4	00	1	18	D
	49	1	56	-	-	T

$M = 1.89 \pm 0.34$ min. 1.67 ± 0.34 min.
$\sigma = 1.73 \pm 0.24$ 1.63 ± 0.24

Results of experiments in mental suggestion of sleeping and awakening where sender and percipient are in the same room

NOTE. The following applies to Table 5 and the ensuing tables:

1. The date column gives the day and month when the experiment was carried out. By comparing dates and Serial Numbers we see that the number of experiments carried out on different days (let's call them sittings) is not always the same.

2. In the column S—H ("go to sleep"—hypnosis) we have the time in minutes and seconds (measured with a stopwatch) which elapsed between the beginning of mental suggestion and the moment at which the subject actually fell into hypnotic sleep. In column W—V ("wake up"—vigilance) the intervals in time are given that separated the beginning of mental suggestion and the moment of awakening.

3. Cases where results are entered only in the Column "S—H," but are absent in "W—V" and vice versa, require an explanation. For example, in the experiment of 19/12 (Table 5) experimental results are entered only in the column of waking (W—V); this means that at this sitting only waking up was effected *mentally*, but that mental suggestion of going to sleep did not take place—the subject was put to sleep *verbally*. At the sitting of 29/12 (Table 5) we have the reverse situation: *mental* induction of sleep, but this was followed by *verbal* awakening. The number of sittings with a given subject can be seen from the dates, and the number of experiments in mental suggestion of sleeping and awakening are entered in the results columns 3 and 4. Thus, columns 1, 2, 3 and 4 show that with Ivanova 16 experiments were conducted in 10 sittings, and what were the results obtained.

4. In the "Sender" column D: Dubrovsky, T: Tomashevsky, V: Vasiliev.

   In order to convey as vividly as possible the conditions in which our experiments were carried out (the relative location of the experimental rooms, the various pieces of equipment, the places of sender, percipient, assistant and so on) we add a plan of our laboratory (Fig. 18).
   The results of the experiments in Group 1 are given in Table 5. As has already been pointed out, in these experiments sender and percipient were in the same round room B [Б in the diagram—see explanation under Fig. 18] at a distance from each other of several metres. The sender was outside the subject's range of sight.
   The statistical evaluation of the numerical data of this table shows that the subject Ivanova fell asleep on the average within 6.25 mins after the beginning of mental suggestion to go to sleep (average error $m \pm 2.06$ minutes) and woke up on an average 2.94 minutes after the beginning of the suggestion to wake up (average error ± 1.13 minutes). In this group of experiments the subject

Fedorova gave the best results: for mental induction of sleep $M = 1.89 \pm 0.34$ minutes; for mental awakening $M = 1.67 \pm 0.34$ minutes.

One weighty objection can be raised against these experiments, namely that in this group of experiments sender and percipient were in the same room, and consequently the experimental results could have been affected by some involuntary verbal murmuring on the part of the sender (in the nature of an ideomotor act) or that some other signals made by the sender could in some way have influenced the subject.

This suspicion, always justified in cases under such conditions, loses its foundation, however, in experiments of Group 2, where the sender was in the round room (B) and the percipient in the small room (A) divided from the first room by two intervening passages (see plan of the laboratory, also explanation of letters) all doors between rooms being carefully closed. The results of these experiments are given in Table 6.

Table 6
Second Group of Experiments

		Subject: Ivanova				
		S—H		W—V		
Date	Serial No.	Min.	Sec.	Min.	Sec.	Sender
29/12	1	9	00	-	-	D
3/1	2	7	00	-	-	D
	3	9	00	-	-	D
11/1	4,5	6	00	2	00	D
15/1	6	1	00	-	-	D
	7	1	00	-	-	D
	8	2	00	-	-	T
21/2	9,10	0	30	1	00	D
3/3	11	4	00	-	-	D
	12	1	00	-	-	D
23/3	13	15	00	-	-	D
27/3	14,15	1	00	8	00	D
	16	9	00	-	-	D
14/4	17	6	00	-	-	D

$M = 4.05 \pm 0.93$ min. 3.67 ± 2.18 min.
$\sigma = 3.50 \pm 0.66$ 3.77 ± 1.54

| | | Subject: Fedorova | | | | |
| | | S—H | | W—V | | |
Date	Serial No.	Min.	Sec.	Min.	Sec.	Sender
22/12	1	-	-	3	00	D
26/12	2,3	5	00	0	10	T
28/12	4,5	1	00	1	00	D
29/12	6,7	7	00	10	00	T
	8,9	1	00	1	00	T
	10	1	00	-	-	T
3/1	11,12	1	30	5	00	T
	13,14	3	30	1	00	T
8/1	15,16	15	00	0	30	T
13/1	17,18	7	00	4	00	T
	19,20	1	00	1	00	T
16/1	21,22	2	00	2	00	T
	23	5	00	-	-	T
25/1	24,25	3	00	2	30	T
	26	0	30	-	-	T
14/2	27,28	4	00	9	00	T
	29,30	2	00	7	00	T
	31,32	3	00	0	30	T
	33	-	-	10	00	D
23/3	34	2	00	-	-	T
27/3	35	9	00	-	-	D

M = 3.87 ± 0.83 min.   3.60 ± 0.88 min.
σ = 3.62 ± 0.59   3.52 ± 0.62

		Subject: E. S.				
26/1	1	-	-	1	00	T
28/1	2	20	00	-	-	T
	3	3	00	-	-	T
31/1	4	0	10	-	-	T
	5	0	10	-	-	T
	6	0	10	-	-	T
22/2	7,8	2	30	7	30	T
	9	2	30	-	-	T

M = 4.07 ± 2.70 min.   4.25 ± 3.24 min.
σ = 7.13 ± 1.91   4.58 ± 2.29

*Results of experiments in mental suggestion of sleeping
and waking, sender and percipient being in different rooms*

The statistical calculation of the numerical data of this table for Ivanova yields M = 4.05 ± 0.93 minutes (time taken to fall asleep) and M = 3.67 ± 2.18

(time taken to wake up). For Fedorova the average time to go to sleep = 3.87 ± 0.83 minutes and the average time for waking up = 0.60 ± 0.88.

Comparing these results with those obtained in the first group, we see that they are somewhat poorer in the case of Fedorova, the mean time taken for going to sleep rose from 1.89 to 3.87 minutes and the time of waking up from 1.67 to 3.60 minutes. For the other subject, Ivanova, on the other hand, the figures in the two cases do not differ substantially from one another.

Let us compare the results of the first and the second series of experiments (by R. A. Fisher's method, described in the preceding Chapter).

It follows 1. the deterioration in Fedorova's experimental results may be considered established; 2. the influence of distance on Ivanova's performance is not proved. (One must note the fact, however, that fewer experiments were carried out with her—see "$n_1 + n_2 - 2$" is equal to the number of degrees of freedom.)

Table 7

Serial No.	S—H			W—V			Subject
	$n_1 + n_2 - 2$	t	P	$n_1 + n_2 - 2$	t	P	
1	20	0.97	0.3 - 0.4	9	0.30	0.7 - 0.8	Ivanova
2	43	2.21	0.02 - 0.05	37	2.05	0.02 - 0.05	Fedorova

Comparison of the results of the experiments in mental suggestion of sleeping and awakening: first and second groups of experiments

Experiments of the third group were conducted under conditions of screening. The sender, at a time unknown to the subject, stepped into the perfect screening chamber which, as may be seen from the plan, was in the round room B. In most of the experiments the subject was in another room (the small room) and only in the same room on a few occasions. The sender from inside the room, when starting mental suggestion ("sleep!" or "wake up!"), unlocked the chain of the register by means of the apparatus already described in Chapter 3.

The results of the experiments are computed in Table 8. The statistical evaluation of the numerical data obtained yields the following results: the average time for induction of sleep for Ivanova = 2.02 ± 0.61 minutes; average time of waking up for the same subject = 3.64 ± 0.81 minutes. For Fedorova the same values are: 3.50 ± 1.35 minutes (in cases of sleep induction), 5.14 ± 4.28 mins. in cases of awakening.

Table 8
Third Group of Experiments

Date	Serial No.	S—H Min.	S—H Sec.	W—V Min.	W—V Sec.	Sender
		Subject: Ivanova				
29/12	1,2	1	00	6	00	D
	3	-	-	3	00	D
23/1	4,5	1	00	0	30	T
	6	-	-	2	00	D
	7	-	-	2	00	D
29/1	8	3	00	-	-	T
26/2	9,10	0	03	5	00	T
	11,12	0	03	1	00	D
	13	0	03	-	-	D
9/3	14	-	-	9	00	T
21/3	15,16	5	00	0	30	D
	17	3	00	-	-	T
29/3	18,19	2	00	7	00	D
	20	-	-	1	00	D
2/4	21	-	-	2	00	D
8/4	22,23	5	00	9	00	T
23/4	24	-	-	3	00	D

$M = 2.02 \pm 0.61$ min. 3.64 ± 0.81 min.
$\sigma = 1.92 \pm 0.43$ 3.03 ± 0.57

Date	Serial No.	S—H Min.	S—H Sec.	W—V Min.	W—V Sec.	Sender
		Subject: Fedorova				
19/12	1,2	0	30	1	45	T
	3,4	0	30	0	30	T
	5	-	-	0	20	T
25/1	6	7	30	-	-	D
29/1	7,8	5	00	18	00	D
1/4	9	4	00	-	-	T

$M = 3.50 \pm 1.35$ min. 5.14 ± 4.28 min.
$\sigma = 3.02 \pm 0.96$ 8.56 ± 3.03

Date	Serial No.	S—H Min.	S—H Sec.	W—V Min.	W—V Sec.	Sender
		Subject: E. S.				
28/1	1,2	0	30	0	30	T
	3	-	-	4	30	T

$M = 2.50 \pm 1.99$ min.
$\sigma = 2.82 \pm 1.41$

Results of experiments in mental suggestion of sleeping
and awakening with screening of the sender

If in the case of the experiments of the third group there had still existed some possible doubt as to whether our chamber adequately excluded radio-waves from the sender's brain (despite the favourable results of the tests of the screening properties of this chamber described in Chapter 3)—in this fourth group of experiments the possibility of the transmission of electromagnetic influences from the sender's to the percipient's brain would seem to have been reduced to a minimum. Nevertheless, these results proved to be the same as those obtained in all preceding experiments: mental induction of sleeping and awakening were observed as before and the speed required for the mental sug-gestion to take effect was not impaired by screening. The quantitative material is given in Table 9.

For Ivanova, as regards mental suggestion to go to sleep, we have an arith-metic mean $M = 5.00 \pm 1.22$ minutes; for waking up $M = 5.38 \pm 1.46$. For Fedorova for "sleep!" the mean was only $M = 0.83 \pm 0.34$ minutes; and for "wake up!" $M = 1.87 \pm 0.41$ minutes. One gets the impression that under con-ditions of double screening the effects of mental suggestion not only failed to deteriorate—they actually improved.

Summarising the results of all these experiments, we arrive at a complete confirmation of the conclusion already referred to in the last Chapter, arrived at in connection with the experiments conducted according to the method of Academician V. F. Mitkevich: the screening used in our experiments, which prevents the diffusion of electromagetic waves to a great extent, and over a wide range of wavelengths, including the Cazzamalli radio waves, does not to the slightest degree affect the transmission of mental suggestion from the brain of the sender to that of the percipient.

Table 9
Fourth Group of Experiments

Subject: Ivanova

Date	Serial No.	S—H Min.	S—H Sec.	W—V Min.	W—V Sec.	Sender
15/4	1,2	2	00	3	00	D
	3	7	00	-	-	D
	4	-	-	6	00	T
19/4	5,6	4	00	2	00	T
	7,8	7	00	4	30	D

$M = 5.00 \pm 1.22$ min. 5.38 ± 1.46 min.
$\sigma = 2.44 \pm 0.86$ 2.91 ± 1.03

Subject: Fedorova

Date	Serial No.	S—H Min.	S—H Sec.	W—V Min.	W—V Sec.	Sender
5/4	1	-	-	0	15	T
9/4	2,3	1	00	1	00	D
	4,5	0	15	1	00	D
	6,7	3	00	1	00	D
	8	-	-	1	00	D
11/4	9,10	0	0	0	0	T
	11	-	-	2	00	T
16/4	12,13	0	0	4	00	T
	14	0	0	-	-	T
19/4	15,16	0	30	0	30	T
	17,18	1	45	4	00	T
22/4	19,20	1	00	2	00	T
	21	-	-	4	00	T
26/4	22	-	-	5	00	V
5/5	23	-	-	1	00	T
9/5	24	-	-	1	15	T

$M = 0.83 \pm 0.34$ min. 1.87 ± 0.41 min.
$\sigma = 1.01 \pm 0.24$ 1.60 ± 0.29

NOTE. 0 (zero) in columns S—H and W—V means that the transmission of mental suggestion in this instance took place with such speed that it could not be measured by means of kymographic registration. In computing the average magnitudes these zero variants were taken into consideration.

*Results of experiments in mental suggestion of sleeping
and awakening with screening of both sender and percipient*

This is apparent from a simple comparison of the numerical data in Tables 5 and 6 (*without* screening) with the corresponding data in Tables 8 and 9 (*with* screening). We will add to this that the kymogram referred to at the beginning of this Chapter (Fig. 19) which gives such striking coincidences between the time of beginning of mental suggestion by the sender with the onset of the response of the recipient, refers to one of the experiments in the third group (where the sender was in a "perfect" screening chamber and the percipient outside it, in the same room).

However, our conclusions emerge with particular force from the statistical computations made subsequently of the quantitative material presented.

By putting together the data given in Tables 5 and 6 and summing up the time taken to send to sleep and to wake up, we obtain a general characteristic of the reactions without screening (see Table 7). A similar procedure applied to the data contained in Tables 8 and 9 yields a general characteristic of the reactions with screening.

As a result we obtained the picture (after Fisher) shown in Table 10. Since the significance criteria for both experimental conditions is $t < 2$, and $P > 0.05$, (it will be remembered that P is the probability of the null hypothesis and the criterion of significance of the difference) there is no basis for attributing to screening any influence on reaction time.

Table 10

$M \pm m$				
Without Screen	With Screen	t	P	Subject
4.28 ± 0.75	3.52 ± 0.52	0.84	0.4	Ivanova
2.60 ± 0.31	2.23 ± 0.64	0.52	0.6	Fedorova

Summary of results of experiments in mental suggestion of sleeping and awakening, all groups of experiments

Thus the statistical analysis of the factual material again supported our basic finding. Contrary to all expectation screening by metal did not cause any even faintly perceptible weakening of telepathic transmission. Even under conditions of double screening mental suggestion continued to act with the same degree of effectiveness as without screening.

Is it necessary to emphasize that such a conclusion may be of tremendous theoretical importance? According to our conclusion the Cazzamalli "brain wireless waves," if in fact they exist, have no connection whatever with the phenomena of mental suggestion. It follows that the energetic factor in telepathic transmission must be sought in quite a different region of the electromagnetic spectrum: either in the region of radiations with a shorter wave (Röntgen or gamma-rays) which is improbable, or alternatively in the region of kilometre waves, or of static electric fields. However, the possibility of these last two factors playing a part also is hardly feasible (see Chapter 9). We fully appreciated the responsibility involved in reaching such a conclusion; and this induced us to undertake a series of different painstaking investigations with the aim of critically evaluating the conditions under which our experiments with the hypnogenic method were carried out, and the empirical results which they yielded. We cast doubt once more upon the correctness of our inferences and we ourselves attempted to refute them.

We will now give a brief account of these control investigations.

7

Critical Evaluation of the Hypnogenic Method and of the Results Obtained by Its Application—Improved Version of the Hypnogenic Method

It must be noted at the outset that the hypnogenic method gave us more permanent positive results than the methods of mental suggestion described in previous chapters. As has been pointed out in the last chapter, only in 6 cases did mental induction of sleep fail to take effect, and mental awakening in 21 instances.

It should be added that in the cases where mental induction of sleeping and waking failed to function there were usually definite reasons: for instance: obviously agitated condition of the subject, the presence of strangers, the presence of some collateral stimulation (such as unusual noises, tightness of garments, etc.). After the removal of such impediments, effective mental suggestion usually did take place. In some cases, apparently, mental awakening was not effected because it was suggested by a different sender from the one who transmitted the suggestion to go to sleep.

The hypnogenic method permits such control over the appearance of mental suggestion that we actually dared, on numerous occasions, to demonstrate—and not without success—this, generally speaking, highly "capricious" phenomenon to visitors admitted to watch our work.

However, when we systematically applied the hypnogenic method we encountered one phenomenon which could have been a serious source of possible mistakes. This phenomenon consisted in the following: after subjects have been put to sleep and awakened again, whether by mental or verbal suggestion, they develop a special form of narcolepsy or auto-hypnosis. The subject begins spontaneously to fall into a hypnotic state and to emerge from it again, and this was not noticed in our first experiments with the same subject. We are here probably dealing with a hypnogenic conditioned reflex which develops when the experiment is repeated in the same surroundings: these surroundings of the experiment become a complex conditioned stimulus or signal causing sleep (see Chapter 6).

Such auto-hypnosis was noticed as far back as the 'twenties, when Academician V. M. Bekhterev was carrying out experiments to induce sleep by mental suggestion jointly with myself on Mikhailova, a subject of Professor K. I. Platonov.

In all, we registered in the experiments with our three subjects 9 instances of spontaneous falling asleep and 18 spontaneous awakenings, which amounts to 10.04 percent of the total number (260) of effective mental inductions of sleep and awaking. By way of illustration of such an "hypnogenic automatism" we reproduce the kymogram taken in the experiment of 25/1/1934 (percipient: Fedorova, sender: Tomashevsky).

As may be seen from a perusal of this kymogram (Fig. 20), at 9:30 the subject stopped compressing the balloon and unexpectedly fell into a hypnotic state (auto-hypnosis). At 9:44 she spontaneously woke up again, which can be seen by a resumption of the swing of the curve. At 9:50 a mental order was given to go to sleep 3 and within 3 minutes the subject fell asleep ("Г," 9:53), and thereafter at 10:15 she again woke up spontaneously. A new mental order—"go to sleep" was obeyed, and immediately after this, at 10:37 1/2 the subject was mentally awakened.

It was imperative to find a means of reducing this phenomenon. First we began to refrain from frequently repeated mental inductions of sleep and waking at a given sitting. Then we noticed that subjects do not fall into auto-hypnosis if an observer is with them all the time and carries on a conversation with them. Similarly subjects already in a hypnotic state do not wake up spontaneously if a conversation is carried on with them.

It is certainly remarkable that such a diversion of the subject's attention which inhibits auto-hypnosis in no way handicaps the effectiveness of mental suggestion to go to sleep and wake up. However, this is only the case if the observer is a person well known to the subject; the presence of outsiders whom the subject hardly knows acts as a strong source of stimulation which will delay, and often prevent, the implementing of mental suggestion. We noticed instances where a subject who had received a mental order to go to sleep did so without finishing a phrase she was in the middle of uttering, and only after being mentally woken up did she complete it.

The measures we took, which have just been described, virtually obviated "hypnogenic automatism," and we were able to satisfy ourselves of this by setting up a number of control experiments. These were carried out as follows: the subject was in the same experimental conditions as usual in our experiments in mental suggestion (i.e. lying on the bed, compressing a balloon etc.). Nevertheless the subject remained awake for an hour or more, answering questions asked from time to time by the observer. But when at last the sender (either the same observer or someone else among those present) began his mental suggestion to her to go to sleep, the subject went into hypnosis within 1 to 5 minutes. She would remain in the hypnotic condition for an hour or more continuing her conversation with the observer, and wake up again within 1 to 3 minutes after the sender gave her the mental order to wake up. Table 11 gives a general summary of these control experiments.

Table 11

			PERIOD OF OBSERVATION						
Serial No.	Location of Subject	Condition of Subject	Beginning H.	Min.	End H.	Min.	Duration H.	Min.	Subject
1	S.R.	H	9	31	10	37	1	06	I.E.M.
2	S.R.	H	11	19	11	49.5	0	30.5	I.E.M.
3	S.R.	H	10	02	11	07.5	1	5.5	I.E.M.
4	S.R.	H	10	45	11	15	0	30	I.E.M.
5	S.R.	V	9	35	10	06	0	31	I.E.M.
6	F.C.	V	10	30	10	43	0	13	F.K.G.
7	F.C.	H	9	56	10	40	0	44	F.K.G.
8	F.C.	V	10	40	10	59	0	19	F.K.G.
9	F.C.	V	9	40	10	50	1	10	F.K.G.
10	F.C.	H	10	50	12	26	1	36	F.K.G.
11	S.R.	V	10	20	12	20	2	00	F.K.G.
12	S.R.	H	9	03	10	02	0	59	F.K.G.

S.R.: Small Room   F.C.: Faraday Chamber   H: Hypnosis   V: Vigilance

Results of control experiments to determine the duration of the conditions of hypnosis and vigilance caused by mental suggestion

The kymogram of one of these experiments is given in Fig. 21. At 9:40 the subject lay down on the couch in the Faraday chamber; she was asked to compress the balloon as usual. The observer (Tomashevsky) was with her. He exerted no mental influence. As can be seen from the kymogram, the subject for a long time continued to compress the balloon without falling into hypnosis. At 10:10 the subject was given a 15 minutes' rest, after which the compression of the balloon was resumed for a further 20 minutes. Finally, at 10:45, the sender (Tomashevsky) began mentally to send her to sleep. As may be seen from the kymogram, the reactions of the subject thereupon began to assume a different character: the swings of the curve gradually flattened out and the intervals between swings grew longer. Finally, in 4 mins. 45 secs. from the beginning of mental suggestion the curve dwindled to nothing and the subject went into hypnosis. Most of the experiments included in the statistical compilations given in the preceding Chapter were carried out under conditions impeding autohypnosis. But in subsequent experiments auto-hypnosis began to manifest itself again. We therefore had to find new subjects who would be suitable for experimentation by means of the hypnogenic method, and such subjects were found. Amongst their number were intellectual and professional people (a teacher, an interpreter and others) and manual labourers; some subjects were normal and healthy, while others were neurotic.

In the second period of our work (1935 to 36) we used these subjects, employing the same methods under the same and also under different conditions to be described below.

The first experiments in mental suggestion carried out with "fresh" ["unsophisticated"] subjects who do not as yet know anything about the experiments nor what to expect are particularly valuable. Here are some such first experiments with 4 new subjects.

Experiment 1. Subject Z. came to the laboratory for the first time in order to participate in an experiment. It was decided there and then immediately to proceed with mental suggestion of sleep to Z. without re-entering the room and without warning her that the experiment was about to begin. Observing the subject through a small window in the wall the experimenters noted that she went to sleep within 1 min. 20 sec. from the beginning of suggestion and from within 3 mins. of leaving her in the room.

Experiment 2. Subject I. arrived at the laboratory accompanied by Dr. Dubrovsky; she entered room A and conversed with him for 20 minutes. The subject did not know of the presence in the laboratory of a second experimenter (the sender). Within 20 minutes from the time of their arrival the sender (Tomashevsky) from room B started on mental suggestion to go to sleep. After 2 mins. from the beginning of suggestion I.'s speech stopped, and sleep set in.

Experiment 3. Subject B. came to the laboratory for the first time to take part in the experiments. At 6:10 she was asked to sit in the arm-chair quietly and to press the air pressure rubber balloon in room B. The compressions of the balloon were registered on the kymographic record in another room, V, in which the sender was (see Fig. 22, for explanation of the letters "B" and "V" see legends to Figs. 18 and 22). At 6:21 the sender began to suggest sleep. After 30 seconds sleep set in. At 6 h. 23 min. 30 sec. the sender began suggesting mental awakening, and at the same instant the registrations began to manifest a state of vigilance.

Experiment 4. Subject K. came to the laboratory for the first time. He was requested to compress the rubber balloon and to stop pressing should he go to sleep. K. was left in room B, the sender in room A (Fig. 22). The subject started compressing the balloon at 8:20. At 8:22 mental suggestion to go to sleep was started, and in 4 minutes the subject fell asleep. After 3 minutes the sender made an attempt at mentally suggesting to the subject that he should wake up (for response to this suggestion see Fig. 19). At 8:32 the sender began to suggest awakening for a second time, and within 1 minute the first swings appeared on the record and within 2 1/2 minutes the subject woke up.

It must be borne in mind that these experiments with our new subjects were carried out when the various pieces of apparatus had been assigned new places (as compared with Fig. 18). The recording appliances were transferred from the round room B into the side room V (see Fig. 22). The round room was now only used by the percipients. The sender was in the most distant room A. In that room a new "perfect" screening chamber was installed (No. 3), made of sheet lead 3 mm in thickness.[1] The sender stepped into the chamber through an opening made in its "roof," a step ladder being used to get to it (Fig. 23). The distance between chambers 1 and 3 was 12 metres. Room A and the round room B were separated by two intervening rooms with three closed doors. Later on, when the experiments of mental induction of sleep were repeated, spontaneous going to sleep as a conditioned reflex manifested itself again in the new subjects. This led us to the working out of the second version of the hypnogenic technique in which falling asleep as a conditioned reflex (or learnt pattern) did not interfere with the programme of the experiments. The essentials of the new version of the experiments were as follows:—

The subject was left alone in the room, and all the time that he was awake he compressed the air transmission balloon. In this variation of the experiment the Mitkevich apparatus with the white and black rotating discs was again used. The sender left the round room B (see plan of the laboratory) where the subject was, went to room V (see plan of laboratory and cf. legend) and started up the kymograph and the rotation apparatus, simultaneously recording with the help of a stop watch the time of the beginning of the experiment. If the white disc appeared, the entire experiment was carried out without any mental suggestion of sleep.

There was continuous kymographic recording to register whether the percipient fell into auto-hypnosis or, possibly in some instances, natural sleep. When the black disc showed on the rotary apparatus the sender immediately proceeded to exert mental suggestion to send the subject to sleep until the curve on the kymograph stopped fluctuating, thus indicating that sleep had set in. The time that elapsed between the beginning of the experiment and the onset of sleep was also noted down after consulting a stop watch.

The value of this experiment from the point of view of method consisted in the following:—

1. The subject was left alone during the entire time of the experiments. He could therefore get no signals from the sender or the observer.

2. The sender himself only learned whether or not he was to do any mental suggesting after leaving the percipient's room and setting in motion the rotary apparatus.

3. Being in conditions in which sleep is the appropriate reaction, the subject sooner or later goes into an auto-hypnotic state. But if mental suggestion really exists, the subject will fall asleep significantly sooner than in experiments carried out in the same experimental conditions but without mental influence.[2]

By means of this technique experiments were carried out with four subjects, in the evenings between 7 and 11 p.m., in electric illumination. 53 experiments in all were conducted; out of these, 26 were done *with*, 27 *without*, mental influencing to go to sleep. The subjects were three women and one man, Z, I, B and K.

The results of these experiments are shown in Table 12, in which the times that elapsed between the beginning of the experiment and the onset of sleep are recorded. In view of the fact that the number of experiments carried out with each subject was not large, and since about the same number of experiments were conducted with each subject it was not deemed necessary, in this instance, to specify which of the results refer to which particular experimental subject.

From statistical computations of the experimental material the following data were obtained: without mental suggestion to go to sleep the average length of time taken to fall asleep turned out to be $M_1 = 17.7$ minutes with an average error of $m_1 = \pm 1.86$ min.; *with* mental influencing the average time taken to fall asleep appeared to be only $M_2 = 6.8$ min. with an average error of $m_2 = \pm 0.54$.

Thus, with mental suggestion, the time taken to go to sleep was almost divided by three. The question arises whether such a difference can be considered reliable, and in order to answer it we applied the formula for calculating the coefficient of probability of the difference

$$t = (M_1 - M_2) / \sqrt{(m_1^2 + m_2^2)} > 2$$

By substituting the figures obtained in our experiments we have

$$t = (17.7 - 6.8) / \sqrt{(1.86^2 + 0.54^2)} = 5.65, \text{ therefore } t \gg 2$$

In the experiments here referred to, the coefficient appeared to be 5.65. Therefore the difference between the results obtained with and without mental influencing may be considered absolutely reliable. For a more graphic representation of the results, we reproduce them here by means of a diagram (Fig. 24).

Table 12

| | Without Suggestion | | | With Suggestion | |
| | Time Interval between Beginning of Experiment and Onset of Sleep | | | Time Interval between Beginning of Experiment and Onset of Sleep | |
Serial No.	Min.	Sec.	Serial No.	Min.	Sec.
1	5	00	1	6	00
2	10	00	2	5	40
3	11	00	3	7	30
4	7	00	4	7	10
5	16	00	5	6	00
6	17	00	6	7	10
7	6	00	7	8	00
8	6	00	8	4	45
9	5	00	9	2	00
10	18	10	10	5	40
11	8	00	11	6	00
12	24	10	12	6	20
13	25	35	13	12	10
14	21	10	14	8	15
15	30	55	15	4	30
16	24	20	16	7	15
17	28	00	17	8	05
18	21	00	18	12	55
19	11	25	19	4	25
20	29	25	20	3	10
21	25	00	21	4	30
22	29	35	22	8	00
23	12	15	23	7	50
24	41	00	24	12	08
25	21	40	25	5	10
26	10	45	26	7	00
27	11	45			

$M_1 = 17.7 \pm 1.86$ min. $M_2 = 6.8 \pm 0.54$ min.

*Results of experiments in accelerating onset of auto-hypnosis
by means of mental suggestion to go to sleep*

Another 10 experiments were carried out under the same conditions with the subject N, 46 years old, absolutely healthy. The results are shown in Table 13.

Table 13

| | Time at Beginning of Experiment | | Time Interval from Beginning of Experiment to Onset of Sleep | | | |
| | | | Without Suggestion | | With Suggestion | |
Serial No.	H.	Sec.	Min.	Sec.	Min.	Sec.
1	8	20	-	-	6	00
2	8	10	5	00	-	-
3	8	10	7	10	-	-
4	5	15	-	-	5	00
5	6	50	-	-	3	15
6	6	55	6	00	-	-
7	7	25	-	-	3	30
8	8	27	8	15	-	-
9	6	45	5	45	-	-
10	7	30	-	-	1	50

M_1 = 6 min. 50 sec. M_2 = 3 min. 55 sec.

Results of experiments with subject N in accelerating the onset of auto-hypnosis by means of mental suggestion to go to sleep

Without mental influencing sleep set in, with this subject, within an average of 6.50 minutes. *With* mental influencing sleep set in within an average of 3 min. 55 sec. As can be seen from these experiments, in the case of N, falling asleep was much quicker with, than without, mental suggestion.

We have put together, in figures, the data obtained in experiments employing the second version of the hypnogenic method (speeding up of falling asleep) for six experimental subjects. The results are given in Table 14.

On the basis of these experiments we arrived at the following conclusion: the process of falling asleep can be accelerated by mental suggestion to go to sleep in instances where the subject tended to go either into an auto-hypnotic state or to fall into natural sleep.

This influence is exerted without participation of any of the receptor organs of the experimental subject. Consequently it is effected by means of mental suggestion.

A final series of experiments, using the same technique, was carried out with the most reliable screening of the sender by means of lead. In order to ensure the impenetrability of the chamber by electromagnetic waves, the gutter around the edges of the opening was filled with mercury before the beginning of the experiment. When the leaden lid, which was raised by a pulley, was lowered, its folded edges were submerged in the mercury to a depth of 4.0 cm (see Fig. 23). Such an arrangement ensured complete absence of cracks between the lid and the chamber itself, thereby creating a more complete

Table 14

No.	Average Time from Beginning of Experiment to Onset of Sleep				Subject
	Without Suggestion		With Suggestion		
	Min.	Sec.	Min.	Sec.	
18	20	05	-	-	F.
17	-	-	10	35	F.
7	20	30	-	-	Z.
7	-	-	7	29	Z.
8	26	40	-	-	A.
8	-	-	7	33	A.
7	12	45	-	-	I.
6	-	-	6	30	I.
5	6	12	-	-	K.
5	-	-	5	11	K.
5	6	50	-	-	N.
5	-	-	3	55	N.

Summary of results of experiments on six subjects in accelerating onset of auto-hypnosis by means of mental suggestion to go to sleep

screening of the inner space of the chamber from radiation than the screening that had been obtained in the iron chamber previously used. Besides, as is well known, lead reduces radiation of shorter wavelength (X-rays and gamma rays). This would have been of importance if, contrary to expectation, it had been established that it is just such rays that are the significant factors in transmitting mental suggestion at a distance from the brain of the sender to that of the percipient.[3]

Inside the chamber there was a chair for the sender and a small shelf-table. The chamber was illuminated by a small lamp connected to an accumulator nearby. A coil with an iron core was fixed to the outside wall of the chamber; if this was switched on the contact appliances in the circuit of the electric signalling lamp inside the chamber were disconnected. The screening properties of the chamber were tested by means of a generator of medium electromagnetic waves (about 50 cm in length), and a corresponding radio receiver, and the tests showed that the chamber was completely impenetrable to these rays.

In order to establish the absorption of gamma rays by the walls of the chamber, measurements were made by means of a radium preparation. The measurements were made with a Gess electrometer, the source of the rays was a radium preparation containing 19.3 mg of radium. On the strength of the results obtained it was concluded that the lead chamber absorbs about 17 per cent of the gamma rays of radium.

The lead chamber was placed in room A. The percipient was in an iron Faraday chamber at a distance of 13 m (No. 1) which stood in the round room,

B, and was separated from room A by three walls (see plan of the laboratory, also legend to Figs. 18 and 22). A pipe from the iron chamber (No. 1) led into the recording room, V (Figs. 18, 22 and legends) and this was connected with a Marey apparatus for the purpose of registering on a kymographic ribbon the compression of the rubber balloon in the hand of the percipient who was lying on the bed. In the recording room there were a kymograph for the registration of the balloon compressions, an electromagnetic recorder by means of which the beginning of mental suggestion to go to sleep could be recorded, and a time recorder which marked the time in minutes on the kymographic ribbon. In addition there was, in the recording room, a stop watch for measuring the time from the beginning of the experiment to the moment when sleep set in.

Both chambers were earthed by connection with the water system. Fedorova, our best percipient, was our experimental subject in this entire series of observations; (she also participated in the experiments described in Chapter 6). The experiment was carried out as follows: Fedorova usually came to the laboratory at about 8 p.m. She had a rest, for about 20 to 25 minutes, during which she talked with the experimenters in the round room B. About 10 minutes before the beginning of the experiment the sender (Tomashevsky) went to room A, in which the lead chamber was, before taking with him *one* of three envelopes in each of which there was a note, prepared in advance by one of the other participants in the experiment. Each note contained one of the following assignments: to suggest sleep from the lead chamber; not to suggest anything; or to suggest from the same room, A, but staying outside the lead chamber. The subject remained with the second participant, who observed the progress of the experiment.

The subject entered the iron chamber, placed herself on the bed, took the rubber balloon and received this instruction from the observer. "Compress the rubber balloon as rhythmically as possible all the time that you are awake. Stop compressing when you feel sleepy or when you fall asleep; when you wake up, start compressing the balloon again." The observer then closed the door of the chamber tightly and went to the recording room, shutting two more doors after him. The subject's chamber was not illuminated and it was completely dark, which contributed to the development of auto-hypnosis.[4]

Having entered the recording room, V, (see Fig. 22 and legend) the observer set in motion the kymograph, switched on the stop watch by pressing the knob, sent a light signal to the sender to the effect that everything was ready for the experiment, and then recorded the beginning of the experiment.

The sender, after the observer had gone to the recording room, descended through the hatch into the lead chamber, closed the hatch, switched on the lamp fed by the accumulator and opened the envelope containing the assignment.

If the instruction was "suggest from the chamber" the sender waited for the signal that all was ready for the experiment, and after that he immediately began mentally suggesting sleep. When he had received the signal that sleep had set in, he opened the hatch, came out of the chamber and showed the observer the assignment that had been in the envelope. This was done for purposes of control, and for making the correct entry into the report of the experiment.

If the note picked out read "suggest from the room," the sender opened the hatch and, upon receiving the signal that the percipient was ready for the experiment, began to suggest to her to go to sleep; he stood on the seat of the chair which was in the chamber, the hatch remaining open—in this way his head protruded from the hatch. If the note picked out read "do not suggest," the sender opened the hatch but refrained from suggesting anything and waited for the signal notifying him of the percipient's state.

In all the above eventualities the following conditions were observed:

1. Until the end of the experiment the experimenter in room V did not know what was the nature of the particular assignment, and merely acted as observer and recorder.

2. The sender only opened the envelope with the assignment when he was in the closed lead chamber.

3. The percipient was left to herself under the same conditions as in all other experiments.

A summary of the results obtained under these conditions is given in Table 15.

In the course of time the leaden wall of the gully became subject to dissolution and the mercury began to seep through into the inside of the chamber. Consequently, in the last two experiments with screening, it became necessary to replace mercury by fine lead filings into which the lid of the hatchway was submerged.

In Fig. 25 two kymographic entries of experiments in this series are shown by way of illustration.

An evaluation of the numerical results obtained in 29 experiments yields the following: 12 experiments were carried out without any mental influencing, showing that the onset of sleep from the beginning of the experiment took, on the average, $M_1 = 7.40$ min. In 10 experiments with mental suggestion under conditions of screening $M_2 = 4.71$, and in 7 experiments with mental influencing but without screening of the sender the average time taken for the subject to fall asleep only was $M_3 = 4.21$ min.

Table 15

Group 1 Without Mental Suggestion			Group 2 With Mental Suggestion from Inside Chamber			Group 3 With Mental Suggestion from Outside Chamber		
Serial No.	Time Interval from Beginning of Experiment to Onset of Sleep		Serial No.	Time Interval from Beginning of Experiment to Onset of Sleep		Serial No.	Time Interval from Beginning of Experiment to Onset of Sleep	
	Min.	Sec.		Min.	Sec.		Min.	Sec.
1	7	10	1	5	10	1	3	50
2	4	15	2	1	25	2	2	15
3	4	20	3	3	40	3	10	00
4	8	10	4	3	40	4	4	00
5	6	10	5	3	55	5	4	30
6	14	20	6	3	20	6	3	50
7	6	05	7	4	15	7	1	05
8	6	10	8	3	05			
9	7	10	9	11	00			
10	6	50	10	7	35			
11	14	00						
12	4	05						

$M_1 = 7.40 \pm 0.98$ min. $M_2 = 4.71 \pm 0.86$ min. $M_3 = 4.21 \pm 1.06$ min.
$\sigma = 3.39 \pm 0.69$ $\sigma = 2.71 \pm 0.60$ $\sigma = 2.82 \pm 0.75$

Results of experiments in speeding up the onset of auto-hypnosis by means of mental suggestion to go to sleep with screening of the sender

A comparison of the average, of the arithmetical mean, of the time of onset of sleep *without* mental suggestion (first group) *with* the onset of sleep with mental suggestion from *inside* the chamber (2nd group), and *with* mental suggestion from *outside* the chamber (3rd group), is given in Table 16.

Table 16

Groups of Experiments	t	P
1st and 2nd	2.07	0.02–0.05
1st and 3rd	2.22	0.02–0.05
2nd and 3rd	0.37	0.7–0.8

Comparison of the results obtained in speeding up the onset of auto-hypnosis by mental suggestion with and without screening

It follows that the difference between the time of onset of sleep with suggestion and the time of onset of sleep without suggestion is proved, and that the screening of the sender by a chamber does not affect mental suggestion.

Fig. 26 gives a diagrammatic representation of this series of experiments. In 6 experiments with screening by means of lead, mental suggestion was also exerted to wake up the subject; in such cases the waking up occurred, on an average, within 1 min. 25 sec.

We thus once again obtained data supporting the results of previous experiments with screening by iron. Screening by iron or lead in the manner in which screening was effected by us does not prevent the diffusion of the supposed waves and radiations that transmit mental suggestion.

| 8

Experiments in Mental Suggestion at Long Distances

A number of doubts can arise after a critical examination of the material presented by us. One of these doubts has already been mentioned in the preceding Chapter, namely: would it not be possible to explain the results we obtained by means of the hypnogenic technique by the supposition that the subject develops conditioned reflexes to go to sleep and wake up again? The very set of conditions under which hypnosis is frequently induced would, sooner or later, become a conditioned hypnogenic stimulus.

Furthermore there may have been in our experiments some additional sound stimuli which could have gained the significance of conditioned signals to go to sleep or wake up (such as the noise made by the lowering of the upper part of the screening chamber, the ticking of the electromagnetic recorder, the inflection of the observer's voice, and so on). As is well known, the sleep-inducing and arousing effects of such conditioned stimuli were studied by the school of Academician I. P. Pavlov, particularly by Dr. B. N. Birman, both on dogs and on hypnotised human subjects.[1] In our experiments the danger of interference by conditioned sound stimuli could have been particularly great because of the hyperaesthesia of hearing often encountered in hypnotised subjects, and it would be wrong to disregard this possible source of error. In order to avoid the formation of hypnogenic conditioned reflexes to chance stimuli in our subjects we took a number of precautionary measures:—

1. The intervals between separate tests of mental suggestion to go to sleep and to wake up in our experiments varied from 1 or 2 minutes to an hour or

more. This eliminated the possibility of the formation of a hypnogenic conditioned reflex to time.

2. In some series of experiments the sender and the percipient were placed in two different rooms far apart from each other, the sender being in the round room, B (Б), the percipient in the small room, A, and vice versa. In many experiments there was no observer.

3. In the series of experiments with double screening the only weak point could have been that both chambers were in the same room (the round room): the noise made by the sender in entering the perfect screening chamber due to the lowering of its upper part could have been heard by the subject and might have served as a conditioned signal to go to sleep or to wake up. However, in point of fact, this factor played no part in our experiments. The subject reacted (by going to sleep or waking up) not at the time when such noises might have been noticed, but only when the sender, who was sitting in the chamber, began to send out mental suggestion—often many minutes after entering the chamber.

In the last series of experiments described in the preceding Chapter, the sender and the percipient were in different rooms, A and B, separated by intervening rooms (see Figs. 18 and 22 and legends). If, however, even after our experiments, one considers the possibility of explaining the results in terms of conditioned reflexes to signals from consciously unperceived subthreshold stimuli[2] one should still take into account the experiments in mental suggestion carried out at great distances between sender and percipient as being more convincing.

The London Society for Psychical Research has recorded numerous cases of spontaneous telepathy where the sender and the percipient were at considerable, sometimes tremendous, distances from one another—for example in England and Australia respectively.[3] One must remember that the reason for starting experiments in mental suggestion in the first place, namely those of Richet, was precisely the occurrence of such spontaneous phenomena at great distances. Warcollier, in referring to the experiments in thought transference carried out jointly by himself and Dr. Osty, emphasised that they obtained results both at small and at considerable distances. I have already mentioned in Chapter 6 that the first experiments over long distances in mental suggestion to go to sleep were described by P. Janet and C. Richet in the '80s of the last century. The experimental subject, Léonie B. in a number of experiments, fell asleep as a result of mental suggestion while at distances of more than 1 km from the sender. Similar experiments carried out at great distances are numerous in the literature. Here we will mention only the more outstanding ones.

At the Third Congress of Psychical Research in 1927 Warcollier[4] reported experiments in which he participated, between New York and Paris in both directions, i.e. at distances of about 6,000 km. In some instances the transmitted tasks referred to a previously selected type of object, for example a page of a book, diagram or picture, a drawing of some object, etc. In other cases the object transmitted was not previously thought of, for example the visual image of an acrobat walking on a horizontal bar which came into the sender's mind. 15 experiments in thought transmission from New York to Paris were carried out. There was coincidence in 5 cases (33 1/3 per cent); 20 experiments were carried out in the opposite direction, 5 coincidences being obtained (25 per cent). Warcollier regards the results obtained as definitely successful experiments in telepathy at great distances. The well known writer Upton Sinclair, in his book "Mental Radio," described numerous experiments in mental transmission of drawings (290 in all), including experiments carried out at distances of 25 to 30 miles.[5]

A large series of experiments at very substantial distances was organised by the Athens Society for Psychical Research in 1925. The experiments were carried out between Athens and Paris (2101 km), Warsaw and Athens (1597 km), and Vienna and Athens (1284 km). The results of these experiments were reported by Dr. Constantinides at the Fourth International Congress of Psychical Research. Geometrical figures, pictures, letters and, less frequently, solid objects, were transmitted. Each group of experimenters consisted of several persons who acted either as senders or as percipients. Sometimes the percipients were put in a hypnotic state in the hope of improving the results. Transmission and reception of mental suggestions were synchronised. Each transmission in either direction was made twice and lasted for 5 minutes with an interval of 5 minutes. Comparing the original transmission objects with the reproductions made by the percipients, it is hard not to agree with Dr. Constantinides that these experiments provide evidence for the possibility of mental influencing at great distances.[6]

According to the assertions of one of the collaborators in our experiments. I. F. Tomashevsky, he was fortunate enough to conduct a number of such experiments with a highly sensitive percipient, V. Krot, a 24-year-old peasant woman. Under the influence of mental suggestion she was able, while outside the sender's range of vision, to fall asleep within 15 to 20 seconds under any circumstances and in various situations. V. Krot did not suffer from narcolepsy. During mentally suggested sleep she would speak, but only maintained verbal contact with the person who had sent her to sleep. V. Krot woke up easily and quickly when Tomashevsky mentally suggested to her that she should. She was completely amnesic after waking up: she did not remember that she had been in a hypnotic state. In addition to complying with mental suggestions of sleeping and waking, the subject also responded to sensory and emotionally tinged suggestions.

Tomashevsky carried out the first experiments with this subject at distances of one to two rooms, and under conditions where the percipient could not know or suspect that she would be experimented with. In other cases the sender was not in the same house, and someone else observed the subject's behaviour. Subsequent experiments at considerable distances were successful. One such experiment was carried out in a park at a distance of about 40 to 45 m: mental suggestion to go to sleep was complied with within a minute.

In our researches in the 'thirties experiments at different distances employing the hypnogenic method were carried out as follows:

The sender left the laboratory and went to another floor of the same building or to another neighbouring house (20 to 50 m), or into the street (walking away from the Institute for Brain Research for distances ranging from 500 to 4,500 m). An observer remained with the subject. In some cases the subject was in another part of the town—on the premises of the Psychotherapeutic Dispensary, i.e. at a distance of 7,700 m from the laboratory of the Institute for Brain Research where the sender was. The subject could not know and indeed did not even suspect that an experiment was being carried out with her, and the observer only knew the time at which suggestion would begin very approximately. These experiments were carried out with Ivanova, already well known to the reader; Tomashevsky and Dubrovsky acted as senders.

Under these conditions which, so it would seem, completely ruled out the possibility of any identifying signals, mental suggestions of sleeping and waking were effected, often with the same speed as if the sender were in the same room as the subject.

The success of these experiments led to our setting up experiments at very great distances—from Sebastopol to Leningrad, about 1,700 km in a straight line.

Prior to leaving for Sebastopol the sender (Tomashevsky) agreed in advance with Dr. Dubrovsky as to the day and hour of the experiments. On July 13th Ivanova came to the Psychotherapeutic Dispensary as usual at 5 o'clock and remained there until 7 P.M. However, during the time which had been agreed for the experiment no telepathic transmission came from Sebastopol because the sender fell ill; he was running a temperature. Ivanova, who was during the entire time under Dr. Dubrovsky's observation, showed no sign of sleepiness.

On another occasion, on July 15th, the same subject again came to the Dispensary at about 10 o'clock (the arranged time). At 10:10 the sender began to exert mental suggestion. At 10:11 P.M. a hypnotic condition of the subject was noted. At 10:40 the sender started mental awakening. Precisely at 10:40, as was subsequently established by a comparison of the experimental reports, the subject came out of hypnosis. It should be added that on days for which experiments were arranged, sender and observer checked their watches by radio with Moscow time.

At the time of mental suggestion the sender was alone on the sea front promenade. It was dark. Dr. Kaialov acted as observer of the subject. He had never previously experimented with Ivanova and he knew neither the purpose of the experiment nor the sender's plans. Kaialov conducted the observations at the suggestion of Dr. Dubrovsky who had taken part in deciding on the days and hours of the experiment.

In these two experiments we unexpectedly obtained a good control: on July 13th there was no mental suggestion, and no symptoms of sleep were noted; on July 15th mental suggestions first of sleep and then of waking were made, and these suggestions were fully complied with.

The results of all these experiments are given in Table 17. In order to understand this table clearly one must take into account that, in experiments 2 to 5 and 7, experiments in mental awakening only were carried out (the subject having been put to sleep by the observer by means of verbal suggestion). In experiments 1, 6 and 12 the sender only put the percipient to sleep, the observer waking her up by verbal suggestion. On July 15th the sender carried out two experiments, 10 and 11; he first mentally put the subject to sleep and then woke her up again mentally.

Table 17

| Date | Serial No. | Distance | Time Taken for Suggestion to Take Effect | | Subject |
			Sleeping	Waking	
9/1/1932	1	25 m	up to 1 min.	-	I.E.M.
15/1/1934	2	25 m	-	1 min.	I.E.M.
29/1/1934	3	500 m	-	6 min.	I.E.M.
25/2/1934	4	4,500 m	-	up to 1 min.	I.E.M.
20/3/1934	5	20 m	-	1 1/2 min.	I.E.M.
21/4/1934	6	7,700 m	up to 2 min.	-	I.E.M.
21/4/1934	7	7,700 m	-	21 min.	I.E.M.
8/5/1934	8,9	7,700 m	3 1/2 min.	4 1/2 min.	I.E.M.
15/7/1934	10,11	1,700 km (Sebastopol to Leningrad)	1 min.	up to 1 min.	I.E.M.
6/1/1935	12	50 m	2 min.	-	A.A.

A total of 12 experiments is given in the table.

Results of experiments in mental suggestion in sleeping and awakening at greater distances

As a control of the subject's condition when she was outside the laboratory, in another set of experiments, a radio set-up was used, working on UKV. The subject was at home where a radio transmitter was installed which had a specially designed switch. As soon as the rubber balloon was compressed the circuit of the current was closed and so caused a generation of radio waves by the transmitter (Fig. 27).

The signals were recorded on the moving tape of a kymograph attached to the receiving apparatus in the laboratory of the Institute for Brain Research. At an arranged time of the day the subject was instructed to plug in her radio transmitter into the electric mains, to take up a comfortable resting position, to compress the rubber balloon and to stop compressing it when sleep set in.

By noting the time from the beginning of the experiment, when the first radio signal appeared, to the time when these signals stopped we were able to observe, at a considerable distance, the time taken for the subject to go to sleep with and without mental suggestion. Not many experiments of this sort were carried out, but the results obtained indicate that the method of using radio signals substantially enhances the experimental possibilities.

Analysing the results of our experiments at various distances, we find that these results are almost identical with those obtained at short distances, from one room to another. The experiment of April 21st, 1934, in which 21 minutes were required for compliance with mental suggestion is an exception.

Natural obstacles such as the curvature of the earth's surface, hills etc. do not affect the phenomena. In this respect "brain radio" does not differ from ordinary radio telegraphy which transmits information from one hemisphere to another; this in no wise contradicts the well known physical law according to which the action of electromagnetic waves is inversely proportional to the square root of the distance between the generator and the receiver of such waves. [Editor's note: That radio transmission attenuation obeys (approximately) an inverse square root law, and not an inverse square law is, of course, due to reflexions at the Heaviside layer. This was unknown to Barrett.]

Only a few authors maintain that an increase in the distance between sender and receiver rapidly weakens the transmission of mental suggestion. Thus the Italian researcher Bozzano writes: "Experiments gave wonderful results when the sender placed his hand on the back of the percipient's head; results are less remarkable when the sender holds the percipient's hand, and still less noticeable when sender and percipient are a certain distance from each other in the same room; results become unstable and unreliable when they are in different places at distances of several hundred metres from one another; finally, when the distance is increased still further a moment comes when experimental results are reduced to nil with the exception of rare occurrences of spontaneous telepathy, which may sometimes intrude into the course of an experiment, but the nature of which is qualitatively different from experimental telepathy."[7]

Our experimental control series involving 5 experimental subjects, set up in the manner described but with the "black or white" method did not support the conclusions reached by Bozzano. We will not here describe these experiments since the material already given in this Chapter obtained by us, using the hypnogenic method, refutes Bozzano's conclusions with sufficient clarity. We will only note that the results obtained by us are of considerable importance for the determination of the energetic nature of the factor that transmits the telepatheme from the sender's brain to that of the percipient. Like the usual radio waves it operates at long distances but, as opposed to radio waves, it is not impeded by metal screening.

If one admits that the "brain radiations" that effect mental suggestion are electromagnetic oscillations, then such oscillations, as has been pointed out by Barrett, must be subject to the "inverse square law," i.e. radiation of spherical waves diffusing in all directions must become weaker proportionately as the square of their distance from their source.[8] Thus, at a distance of one thousand kilometres from the sender the telepathic influence on the percipient must be a million times weaker than the same influence on the percipient when he is at a distance of one metre. It would follow from this that the transmission of the waves, which we have admitted for argument's sake, over a great distance, requires a tremendous output of energy at the origin of the radiation, i.e. the active brain, otherwise these radiations would be so attenuated that the percipient would be incapable of responding to them. Experiments, however, provide reasons for supposing that a similar mental effort is required from the sender in long as in short distance experiments.

Contemporary radio technological achievements provide examples of a different nature. Negligibly small amounts of energy received by radio receiver from a low power generator can be amplified to an enormous extent, and it is consequently possible to receive accurate information over tens of thousands of kilometres. Therefore Barrett's reference to the inverse square law has little relevance for ordinary radio or for telepathic transmission. We will discuss this important question in the last Chapter (10) of this book.

| 9

Some Psychological Peculiarities of Mental Suggestion

The following difficulty presented itself in the course of our research: how does it come about that in many experiments the realisation of mental suggestion of sleeping and awakening occurs, not at once after the onset of suggestion but after the lapse of some time, within two or three minutes or more? In order to solve this question we traced curves generally accepted in variation statistics of the allocation of the variant. Along the abscissa were plotted the time in minutes of the speed with which mental suggestion to go to sleep (Fig. 28A) and to wake up (Fig. 28Б) took effect. On the ordinate were marked the number of fallings asleep and wakings up noted at a given speed. The subjects were Fedorova, Ivanova and E. S.

The graphs here presented show clearly that, in the majority of cases, falling asleep and waking up occurred even before the lapse of one minute from the beginning of the experiment. However, in isolated cases the realisation of suggestion was delayed by 10 or more minutes.

Other graphs, which are not here reproduced, prepared on the same lines, showed that such delays of telepathic effect were apparently dependent to some extent on the qualities of the sender, namely his ability to concentrate on the suggestion that was being transmitted. Thus, in the majority of the experiments in which Tomashevsky was the sender the subject went to sleep or woke up in a short period of time, 1.5 to 1 min., whereas these intervals were longer when Dubrovsky was the sender.

The reason for these delays became partly apparent when we began to employ, in the course of our experiments, the method of questioning our subjects after the manner usual in what is known as "hypnoanalysis." This consisted in questioning the subject as to who had sent her to sleep with the object of eliciting her subjective experiences. Questions were asked such as "Why did you go to sleep?"—"Couldn't you keep awake?"—"Who made you feel like that?"—"What are you thinking about?" etc.

By the use of this method we were able to establish the existence in our subjects of a certain resistance, some sort of negativism, towards the realisation of the mental suggestion already subconsciously perceived. By way of illustration we will quote the report of a characteristic experiment of this type.

Experiment 22. 4/1934. Subject: Fedorova. Sender: Tomashevsky. Observer: Dubrovsky. The subject came to the laboratory at 8:30 P.M. and lay down on the bed in the Faraday chamber at 8:55. The observer was in the chamber with the subject and carried on a conversation with her; the door of the chamber was open.

At 9:12 o'clock the sender, who was in the same room but outside the subject's range of view, began mentally to suggest to her that she should go to sleep. At 9 hrs. 12 mins. 30 sec., after as little as 30 seconds, the subject went into hypnosis and remained in that state for two hours, conversing with the observer. At 11:30 the sender entered the "perfect" screening chamber which was in the same room and began mental suggestion of awakening. At the same time the observer recorded the following verbal statements of the subject: "Joseph Franzevich [Tomashevsky—L. L. V.] says 'wake up'—but meanwhile wind up the ball. I want to rest—sit up, sit up—oh, how he shouts!—what a voice—don't worry, I will soon get up—you will strain yourself!—a pity, your throat—that's enough, do you hear?—tar, tar,—worse than a bitter turnip— put down the receiver—the ball is wound up—that's enough, I am getting up!" While pronouncing these last words, at 11:41, the subject woke up.

At 11:44 the sender starts again to suggest to her to go to sleep. The subject says: "Joseph Franzevich slowly unwinds the thread, orders to go to sleep— enough—hang up the receiver." After a minute at 11:45: hypnosis. At 11:50 the sender starts mentally to wake her up again, all the while remaining in the "perfect" screening chamber. The subject instantly announces: "What's that!— I am fed up with him—he won't let me rest in peace—Joseph Franzevich has an interesting face—sits there with his eyes closed and says—too loud—carefully—wind up the ball." At the last words at 11:54 the subject wakes up.

We have quoted the report of this typical experiment to introduce the reader, after the abstractions and figures of preceding Chapters, to the concrete realities of our everyday observations. Experiments of this nature abound with details which merit attention but which we cannot discuss here. Let us only note some of the features common to the majority of the experiments carried out by us.

1. The impression is created that, although mental suggestion to go to sleep and wake up are immediately perceived by the subject, the implementation

of the perceived suggestion is delayed owning to an initial conscious or unconscious resentment. It should be noted that a similar resentment against the hypnotist's order is often manifested in ordinary (verbal) suggestion.

2. Questioning elicits that the percipient subjectively perceives a connection of some sort with the sender, sometimes symbolised by "a thread," sometimes "the unwinding of a reel," etc. Often mental suggestion is perceived as an order transmitted by telephone. Such details of course cannot give us an understanding of the nature of the energetic influence of the sender on the percipient, but from a psychological point of view such details deserve attention.

3. Questioning elicits that the percipient not only subjectively perceives a connection of some sort with the sender but also recognises which of the experimenters is acting on her by mental suggestion. In this connection one of the experiments carried out by us is most illuminating. We quote the report in full.

Experiment of 20/4/1934. The subject, Fedorova, arrived at the laboratory at 9:30, and she was led to room A for a rest. Prof. Vasiliev then went into the round room B (see Figs. 18 and 22 and legends) where he was joined by Tomashevsky who had previously been with the subject. In the discussion of the programme for the experiment the following plan was made: the subject was to be put into chamber No. 1 which was in room B and Tomashevsky was to remain with her as observer. Vasiliev, who had never before sent the subject to sleep, either mentally or verbally, would pretend to leave the laboratory but would in fact go to the distant room A where he would begin mental suggestion at a moment unknown to the observer (Tomashevsky).

This plan was put into effect. At 9:55 the subject was led out of room A into room B and entered the chamber together with Tomashevsky. Vasiliev staged his exit from the laboratory and went into the distant room A. The observer questioned the subject during the entire time of the experiment, entering his questions and her answers in his report. At 9:58 the sender (Vasiliev) began mental suggestion to go to sleep. At 10 o'clock the subject went into hypnosis.

Subject: No more is necessary.
Observer: Who put you to sleep?
Subject: You. To-day he is good at putting to sleep.
Observer: Who put you to sleep?
Subject: Tomashevsky.
Observer: What else comes to mind?
Subject: Vasiliev is creeping into my head—he came into my mind and now he is creeping into my head.

At 10:18 an idea came to Vasiliev to suggest to the subject the image of a bird which he visualised as a condor or a griffin. At approximately the same time the observer put the following question:

Observer: Tell me what comes into your head.

Subject: He shows well. . . .

Observer: Who is he?

Subject: Vasiliev. His eyes bulge. . . .

Immediately after that:

Subject: A cock—now I see it, he is sitting at a table, a round one. (The agent was actually sitting at a round table.) It was he who took everything from me—

Observer: Who hypnotised you?

Subject: He did it—he paralysed me.

At 10:35 the sender enters the room of the subject and goes into screening chamber No. 2 (Fig. 22). At 10:40 he begins mentally to wake up the subject.

Subject: Stay there for a bit. He is winding up the reel. Enough of it. Professor Vasiliev, stop it!—I shall have to wake up—I don't want to—well, enough—

At 10:40:

Subject: I am fed up with it—Vasiliev is there (points in the direction of the screening chamber)—how he is straining himself!—poor thing! I did hear him.

At 10:43 the subject wakes up. At 10:43.5 the sender proceeds with a second set of suggestions to go to sleep, immediately following the waking up (contrary to usual procedure).

Subject: Something is wrong—

At 10:46 she falls asleep.

It seems to me that the most essential points in this experiment are that the subject not only recognised the sender in spite of the fact that in the two years of their acquaintance he never attempted to send her to sleep, and also that during the entire course of the experiment the subject accurately described the sender's behaviour and surroundings.

These phenomena could either be explained by an ability of the percipient to recognise the sender from a perception of the quality of the energetic influence, or else from an ability to perceive the contents of the subconscious realm of the observer, although in the given case he was at that time far from consciously suggesting anything to her, being fully occupied in questioning her and writing up his report. If the second alternative is correct, we are here dealing with the phenomenon studied in detail especially by French authors, known as "thought reading" (lecture de pensée—Osty and others).[1]

We thus see that the "hypnogenic" method devised by us is not only most useful for carrying out psychophysical experiments, but also permits us to penetrate more deeply into the psychological mechanisms on which mental suggestion depends and a knowledge of which is in any case indispensable for the solution of the problem of the energetic nature of these phenomena.

We must touch on one further question which is but little explored in the current literature—the problem of the so-called direction of telepathic transmission. Analysing this question we reached the conclusion that it is necessary to distinguish between three types of "direction":—

1. Is it essential for the success of a telepathic experiment that the sender should know in which direction with respect to himself the percipient is located? This we agreed to call the "geometrical set" of suggestion.

2. Is it necessary that the sender should know the physical surroundings of the percipient during the experiment? This we provisionally called the "physical set."

3. Must the sender know who is the percipient in a given experiment? Is it necessary that the sender should connect the suggested ideas with the mentally visualised image of the percipient? This we called the "psychological set" of suggestion.

In order to ascertain the significance of each of these three types of "set" we carried out special experiments and obtained the following results: It would seem that both "geometrical" and "physical" set of suggestion have no significance. A number of experiments we carried out by the hypnogenic method show that there is no need for the sender to know where and in what direction the percipient is with respect to himself; the mental putting to sleep and awakening is effective in the usual way without any such knowledge. We will give only one such experiment by way of illustration.

Experiment of 5/4/1934. The subject, Fedorova, came to the laboratory of the Institute for Brain Research at 8:25 P.M. Tomashevsky met her and she lay down on the couch in the small room. It was decided in advance that the sender would be Dr. Kaialov who during the experiment was in the building of the Psychotherapeutic Department in Strike Street at a distance of about 7,770 m from the laboratory. The sender did not know the location of the laboratory and had never been there; he did not know Tomashevsky who was acting as observer, and was with the subject. Fedorova was one of Dr. Kaialov's patients and he was aware that she was the percipient in the experiment.

The watches of sender and observer were compared in advance by telephone.

At 8:50 Dr. Kaialov began mental suggestion to go to sleep. At that same moment (8:50) the subject said: "I must write to Kaialov. But now I won't write after all."

At 9:00 the subject says: "Somehow or another it is peaceful. I don't want to move. My face feels as if it wasn't mine."

At 9:03: "I would like to close my eyes." She closes her eyes, and at 9:04 goes into hypnosis.

At 9:30 Kaialov begins to send suggestions for her to wake up.

At 9:40 the subject announces: "I feel I must open my eyes." She opens her eyes and stares at one point, but at 9:44 she again closes her eyes.

At 9:45 the subject again declares: "Strange, I want to wake up again" and while saying so she wakes up completely.

But if "geometrical" and "physical" set do not play any substantial parts, as can clearly be seen from the above experiment, our experiments indicate quite the reverse as regards "psychological" set: the sender must know who is the percipient in a particular experiment, otherwise mental suggestion remains without effect. Out of a considerable total number of experiments substantiating this point we will here cite the following two:

Experiment of 14/4/1934. The subject, Ivanova, came to the laboratory at 9 P.M. She went into the round room B (see plan of the laboratory and legend) and made herself comfortable in the armchair by the fire. At 9:30 P.M. the second experimental subject, Fedorova, arrived at the laboratory and she was asked to lie down on the couch in the small room A.

Tomashevsky was with her in the role of observer.

At 10:30 the sender, Dubrovsky, entered the Faraday chamber which was in the round room, where the first subject, Ivanova, was.

At 10:35 the sender began mental suggestion to go to sleep, directing it towards Fedorova.

At 11:00 Fedorova went into hypnosis. Between these times, i.e. 10:35 and 11:00, Ivanova three times gave indications of being sleepy, but this did not result in hypnosis and instantly disappeared when she was questioned.

At 11:30 Ivanova was transferred into the small room and put in the armchair, separated by a screen from Fedorova who remained in the hypnotic state. Tomashevsky, the observer, stayed with both subjects.

At 11:33 the sender, Dubrovsky, who was in the round room in the Faraday chamber, began mentally to suggest to Ivanova to go to sleep.

At 11:39 the suggestion took effect—Ivanova went into hypnosis. Fedorova continued to sleep.

At 12:12 the sender, while remaining in the Faraday chamber, begins to wake up Fedorova by mental suggestion.

At 12:25 Fedorova declares: "He is trying to wake me up but nothing happens. Arseny Vassilievich (Christian names of sender) is trying hard." However, the sender did not succeed in waking up the subject. The remainder of the experiment is here omitted.

The following experiment was still more impressive and was kymographically recorded.

Experiment of 19/4/1934. The subjects, Ivanova and Fedorova, were in different rooms of the laboratory, Fedorova being placed in the Faraday chamber. Neither knew of the other one's presence. Fedorova's state was recorded kymographically, Dr. Dubrovsky carried out the observations on Ivanova. The sender, Tomashevsky, was in the round room, i.e. in the same room as Fedorova.

At 19:19 mental suggestion started, aimed at putting Fedorova to sleep. Sleep set in at 9:20. At 10:35 the sender joined Fedorova in the Faraday chamber. At 10:37 mental suggestion was started to put Ivanova to sleep. She fell asleep at 10:41.

At 10:58 mental awakening of Ivanova was begun and at 11:00 she woke up. Meanwhile the subject Fedorova, despite the direct proximity of the sender (Tomashevsky), continued in her sleep state without showing any response to the suggestion to wake up that was directed, not towards her but towards Ivanova.

Similar experiments were carried out on two subjects in three sessions and gave the same results. A response to mental suggestion was elicited in the very subject towards whom the suggestion was directed.

In order to confirm these results some further experiments were carried out by means of the following technique the subject was, as usual, put into hypnosis by means of mental suggestion. She was left in this state for 20 to 30 minutes. During that time the sender proceeded to awaken by mental suggestion some absent person "X," continuing thus for 5 minutes. The sender then proceeded to exert mental influence on the subject. In 4 instances the percipient was within the sender's range of sight and in three instances she was in another room. The results of these experiments are shown in Table 18.

Table 18

Serial No.	Time at Beginning of Experiment		Subject Woke Up After		Location of Subject
	H	Min.	Min.	Sec.	
1	8	39	0	05	From another room
2	8	48	1	15	In sender's range of view
3	5	40	1	00	In sender's range of view
4	7	20	0	30	From another room
5	8	10	0	45	In sender's range of view
6	8	48	0	15	In sender's range of view
7	11	20	2	10	From another room

Results of experiments in awakening a subject by directed after non-directed mental suggestion

As can be seen from this table, in 6 experiments with directed influence awakening occurred in less than two minutes and only in one instance were 2 min. 10 sec. required.

In the control experiments (suggestion to "X") a five-minute period of suggestion failed to wake up the subject. On the basis of the experiments here described and many similar ones we reached a conclusion which, from a psychophysical point of view, may seem somewhat unexpected and difficult to explain, i.e. that "psychological set" would appear to be a necessary condition for mental suggestion. The contents of the transmitted suggestion must be accompanied by a mental image of the person to whom the suggestion is made.

It should be noted that some previous investigators into telepathic phenomena reached similar conclusions. Thus, for example, Caslian wrote concerning telepathic phenomena as follows:

"In order to obtain the best results at telepathic ending it is imperative to visualise mentally the assignment one wishes to transmit and then consciously to send it to the experimental subject while representing his image to oneself as clearly as possible."[2]

Caslian regards these conditions as a factor improving the results of telepathic transmission, but not as imperative. Warcollier, as has already been mentioned in Chapter 8, when conducting experiments from 1923–1925 involving distances from Paris to New York, was convinced that only 5 out of 20 experiments in mental suggestion were successful, transmission being to American percipients from French senders. Where, mentioning this, Warcollier dismisses the possibility that his results might have been obtained by chance, he explains the positive results in this case as being due to *contagion mentale*.

An experiment carried out by Lodge supports the contention that in the absence of a selective set there is no positive effect. This experiment was conducted with a view to determining whether telepathy between persons unacquainted with each other is possible. For this purpose he made use of a long-range transmitting radio station whose listeners were requested to note at a certain time what they were all thinking about. In three out of five series of experiments the general class of the target transmitted was notified, in the remaining series this was not stated. About 25,000 answers were received out of which only a few were approximately correct. The results obtained did not exceed what could have been expected by the laws of chance. An experiment of the same type was organised by G. Murphy in Chicago on March 3rd, 1924. More than 2,000 answers were received and, as Warcollier notes, none of them suggested that there was any telepathic transmission.

Thus, according to the data obtained in our experiments, a two-way and not a one-way connection is established between sender and percipient: on the one hand the sender must direct his transmission precisely at the given percipient, on the other the percipient learns the telepatheme, as well as who is the sender at that moment.[3]

These results of ours find support in the investigations of previous authors. The London researchers reporting on the phenomena of spontaneous telepathy have long ago described cases in which the percipient telepathically perceived experiences of the sender; and the sender, at the same time, also telepathically perceived the situation of the percipient at that moment. These very rare cases were called "reciprocal" (two-way) telepathy.

S. I. Turligin, a pupil of Academician P. P. Lazarev and a Moscow physicist, observed something similar in his experiments. In his work "Radiation of microwaves ($\lambda \approx 2mm$) by the human organism" he writes: "The subject T. N. declared to us that she 'feels' quite well whether the sender is 'working'[4] and can specify the time when he is working. In order to verify this statement we conducted some experiments as follows: The sender was in the chamber, the pipe in its front wall being covered with cigarette paper, and the percipient was sitting on a chair at a distance of 2 m in front of the pipe. We proceeded with the experiment, continuing until the time when the percipient made the first mistake. The signal 'work' or 'do not work' was given to the sender noiselessly by means of pulling a thread. At the beginning the subject proceeded satisfactorily and gave quick and faultless answers as to whether the sender was active or not. But as time went on the answers were generally slowed down by fatigue. In a control experiment a correct answer was given in the 35th minute, and this was the 17th consecutive correct reply.

"The probability that the subject's answer should, 17 times in succession, coincide with the quite arbitrary series of instructions 'work' and 'do not work' by chance alone is

$$\eta = 1/17! = 1/3.55 \times 10^{15} = 2.8 \times 10^{-16}.$$

Such a sequence would be quite unbelievable: there is one positive chance in three and a half milliards of millions of negative chances."[5]

Summarising all that has been said in the last three Chapters, we arrive at the following inferences:—

In the first place, we must conclude that a preliminary selection of sensitive subjects is requisite in experiments for investigating the psychophysical nature of telepathic phenomena. This is necessary in order to obtain sufficiently reliable results.

Secondly, we note that the "hypnogenic" method worked out by us, with self-observation by the subject, coupled with objective registration of his or her responses, is a perfectly suitable method for use in further experiments because the results obtained by this method are clear and repeatable.

Thirdly, we come to the conclusion that it is not necessary for the sender to know either the location or the nature of the surroundings of the percipient. On the other hand, he has to know the subject by sight and must clearly visualise his visual image in order to effect telepathic transmission.

In the fourth instance, it must apparently be accepted that the length of the distance between sender and percipient plays no appreciable part. This does not mean, however, that the inverse square law does not apply to such cases.[6]

Fifth we note that so far no one has succeeded in discovering any physical indicator or radiation produced by the brain which transmits the telepatheme. Cazzamalli's claims and experimental recordings of radio brainwaves were not confirmed in our work.

Finally and sixth we find, contrary to some data to be found in the literature, that screening of the sender from the percipient by means of metal does not prevent the occurrence of telepathic phenomena. From this it must be concluded that, if the transmission of thought at a distance is effected by radiation of electromagnetic energy emanating from the central nervous system, then such electromagnetic energy must either be sought in the region of kilometre electromagnetic waves, or else beyond the soft X-rays, but neither supposition is at all probable.[7]

Such are our inferences. As can be seen, the problem of further researches is becoming more complicated and passes from the purely psychological realm into that of psychophysical experiment, which must either support the hypothesis of the electromagnetic nature of the energy transmitting the telepatheme, or else refute it; the problem then resolves itself into the search for a form of energy the properties of which are as yet unknown and specific to substances of such an advanced nature as the human brain.[8]

In February 1961 the author received a most important communication from the most eminent authority on hypnotism in our country, Professor Constantine Ivanovich Platonov, author of the monograph "The word as a physiological and as a medical factor,"[9] a work that has had three Russian editions and has been translated into foreign languages.

The material sent by Prof. Platonov (see Appendices E and F) confirms that, as far back as 1924, a group of Kharkov scientists carried out a most valuable though not extensive research programme employing the hypnogenic method of mental suggestion and screening of the sender from the percipient. This shorter work represents, so to speak, an abridged version of our investigations over many years. We find with great satisfaction that the results and conclusions given by the Kharkov researchers coincide in many respects with what we discovered ten years later. This similarity in itself is most significant.

| 10

The Present State of the Problem of Mental Suggestion

The investigations described in the foregoing Chapters were made a quarter of a century ago. It is natural to ask the question: what has been done in the subjects touched upon in our researches in the nineteen forties and 'fifties, and up to the present moment? And, first of all; can one at present regard mental suggestion as a definitely and irrevocably established fact? How does the contemporary world of science regard telepathy and other related parapsychical phenomena?

It should be mentioned here that the Second World War brought about great changes in the international organisation of parapsychological research. The International Committee organised by Vett ceased to exist, and the congresses convened by this Committee likewise ceased to be held. The last of these took place in Oslo in 1935. During the war years the Societies and Commissions for parapsychological research in a number of European countries also ceased to exist. After the Second World War the organisational centre for parapsychological studies shifted from Europe to the United States of America. In 1951 an international parapsychological organisation was founded in New York (Parapsychology Foundation Inc.); this established connections with many local centres of parapsychic research which now exist in all five continents.[1] After the war period a number of new societies of this type were founded—in India, Japan, in the South African Republic, in Finland and in many other countries.

This Foundation issues to the world every two months a bulletin containing information as to what has happened in the field of parapsychology throughout the world during that period (Newsletter of the Parapsychology Foundation Inc.). A number of monographs on various parapsychological questions have been published.[2] In 1953 the first post-war Conference of Parapsychological Studies was held at Utrecht, Holland, with the participation of this Foundation. Conferences and symposia followed. Of these mention should be made (as being the more serious ones) of the Conference on spontaneous parapsychological phenomena in Cambridge (1955), and the International Symposium on psychology and parapsychology in Royaumont, France (1956).

As regards mental suggestion, the most important centres of research are at present considered to be: 1. The Parapsychology Laboratory at Duke University, Durham, N.C., U.S.A., 2. the research group of Dr. Soal, London,[3] 3. the Paris Institut Métapsychique directed until recently by the late R. Warcollier. In the last few years research into mental suggestion and related phenomena has been resumed in countries of the People's Democracies; Drs. Ryzl and Figar in Czechoslovakia, Prof. S. Manczarsky and his collaborators in Poland and Dr. M. and G. Jun in the German Democratic Republic are studying these topics.

Every year more and more researchers are becoming convinced of the real existence of mental suggestion and study the various aspects of these complex phenomena. This does not, however, mean that mental suggestion has yet found general recognition by scientists—not in the least! Arguments for and against are still going on. Thus, for example, the researches over many years of Dr. S. G. Soal, which had for long been regarded as being first class work, have now been subjected to the severest criticism in the pages of authoritative journals.[4] Sceptics are also endeavouring to find fault with the quite exceptional experiments of Brugmans[5] mentioned in Chapter 4. Recently the American chemist, Dr. George Price, in his article "Science and the Supernatural"[6] stated that, up to the present day, the published works on this subject are open to criticisms of the experimental technique employed. He is not even satisfied with the statistical calculations of the quantitative data obtained in the experiments—calculations which had, in many cases, indicated a high probability of the existence of mental suggestion.

Thus, for instance, one of Dr. Rhine's subjects in experiments involving guessing the right one out of five symbols in a pack of Zener cards, made 25 correct guesses in succession. This result has a chance of coming about by chance of 1 in 5^{25}, i.e. 1 : 298,023,223,876,953,125. What does this mean? It means that such a case (25 successive correct guesses) would occur, on probability theory, only once in the above mentioned "astronomical" number of tests.[7]

Price begins his article by acknowledging the success of parapsychology. He writes: "It is remarkable that during the past 15 years [1940–1955— L. L. V.] there has hardly appeared one scientific article containing any criticism of the work of parapsychologists. This victory is the result of careful experiment and wise reasoning. The best experiments of Rhine and Soal show far higher results than would be expected by chance alone. The possibility of prompting by means of sensory cues is virtually excluded since the cards and the percipient are often in different buildings. Dozens of experiments have yielded positive results. . . . and the mathematical calculations were approved by investigating statisticians."[8]

One might think: Well, what more could anyone ask? Price is not satisfied with what he himself has said. He sets himself the in my opinion unjustified task of persuading his readers that mental suggestion, and other connected parapsychological phenomena, are "incompatible with modern science"; that they are in conflict with the categories of space, time and causality, and that in consequence admitting them is tantamount to admitting the miraculous. He continues to expand this train of thought by writing: "If parapsychology and modern science are incompatible, why not renounce parapsychology? We do know that the hypothesis that some people lie and deceive themselves fits in well with the framework of science."

The alternative, then, lies between believing in something "truly revolutionary," "radically at variance with scientific thought" on the one hand, and believing in the possibility of premeditated fraud and self-deception on the other. Which seems more reasonable? Price chooses the second alternative, thereby casting a slur on the integrity of parapsychologists: they either deceive gullible readers, or else they themselves are the dupes of the fraudulent machinations of their experimental subjects.

One can only marvel that such arguments should have been brought forward by Price in this day and age when we are witnessing such immense revolutionary and radical changes in physics. Were not quantum and relativity theory, and the denial of the absolute meaningfulness of the principle of the conservation of mass, volume and energy at first "incompatible with the present theories of science?" Can one deny a priori the possibility of such "incompatibles" in the field of psychophysiology, which is a considerably less fully explored, and an immeasurably more complicated branch of knowledge than physics? Price refuses to recognise the simple truth, writ large by the whole history of science and especially during the last decade: that which yesterday appeared "supernatural" is to-day accepted as "natural."

A critic might discredit the results of parapsychological research, say those of Soal and Rhine, in two ways, (1) by finding in these experiments defects of technique which could enable the percipient to guess the intended assignment; (2) by proving fraud to have been committed by participants in the experiment—some collusion, secret signals, etc. Price, of course, does not allege that Soal, Rhine or some other participant in their experiments simply perpetrated some fraud, and he has no proof to that effect. He only asserts that the experiments were so designed that the experimenters, e.g. Soal, could, if he had wanted to do so, have forged the results, and that therefore better proof is required than the data provided by Soal before one could be expected to "believe" in the "supernatural."

Price goes on to describe the set-up of Soal's experiments, in which usually three persons took part; the experimenter was with the sender who mentally suggested the assignment to the percipient in an adjoining room. Price analyses six possible combinations ("procedures") of conspiracy between the sender and the percipient or the experimenter himself, which would be effective in securing the success of the experiment. If, so Price warns his readers, there is collusion between the participants in an experiment, it is extremely difficult to guard against possible normal means of communication. For example, the sender can signal to one of those present by means of some barely noticeable movement, and this observer could make use of a pocket radio generator to transmit the information to the percipient. In short, when faced with the alternatives "fraud or 'miracle'" it is more reasonable to choose fraud.

In concluding his article Price states that he would have been satisfied by only a single positive experiment in which "an error or fraud would be as impossible as the supernatural is impossible."[9]

We have dwelt on Price's critique at some length because it is characteristic of those who, at the present time, deny the possibility of mental suggestion.[10]

However, Price might perhaps not be quite so intransigent if it was only mental suggestion that was in question.

It is typical of the present state of parapsychology abroad [i.e. in the West] that telepathy, clairvoyance and precognition are combined into a single concept, "extrasensory perception," abbreviated as ESP, or still more briefly *psi* (or Ψ, a letter of the Greek alphabet). Not so long ago, before the 'forties, mental suggestion and telepathy were regarded as the simplest and most frequent parapsychological phenomena. Nowadays, however, some parapsychologists, e.g. Rhine, have given up this point of view and have come to regard telaesthesia—the perception of objects without the participation of any of the known organs of sense—as being the basic factor, and that most often encountered in experiments.[11] Telæsthesia is supposedly constantly to interfere in experiments in mental suggestion and to eclipse it.

For example, let us suppose that I mentally suggest to a percipient some picture or object which I am fixating visually and that the experiment succeeds: the percipient draws the same object. If this is neither the result of some involuntary prompting nor of chance, what is it: telepathy or telæsthesia? It used to be assumed that this was telepathy; but modern parapsychologists would assert that it was much more likely to be telæsthesia, considering telæsthesia as a reliably established fact. Experiments in "pure telepathy" are now designed so as to exclude the possibility of "telæsthesia."[12]

It should be noted that, in this connection, the hypnogenic technique of mental suggestion has certain advantages as compared with sensory techniques: in practice it is easier to exclude the supposed admixture of telæsthesia.

Soal's experiments in which, according to his view, "precognition" was involved caused greater alarm on the part of the critics, notably G. Price. Dr. Soal and his collaborator, Mrs. K. M. Goldney, encountered this phenomenon at the beginning of the 'forties in a series of experiments with one of their subjects—Basil Shackleton. In these experiments Shackleton (a nervous and artistic man) had the assignment of telepathically perceiving cards[13] at which the sender in an adjoining room looked in a definite rhythm (2.6 sec.). Positive results were obtained only with three persons who tried their hand as senders. An analysis and a statistical evaluation of the results showed that Shackleton frequently named, not only the card at which the sender was looking at that moment (the "target card" in Soal's terminology), but the one that preceded it ("the card - 1") or the one that followed it ("the card + 1").[14] The first case could have been an instance of the usual mental suggestion, in the second case a trace of the preceding card which remained in the sender's memory might have been transmitted, but how is one to account for the third case? Here neither the conscious nor the subconscious mind (latent memory) of the sender could have been the percipient's source of information. Sometimes without Shackleton's knowledge other experiments were included in the series, in which the sender merely touched the reverse side of the cards without looking at them: in those instances inconclusive results were obtained. Finally, chance results were even more unambiguous when an automatic machine instead of a sender was used to deal the cards.

Soal called statistically reliable higher-than-chance guessing of a card that had not yet been pulled out of the deck ("cards + 1") "precognitive telepathy." When describing these experiments the authors note the part played by the sender. The best senders (as opposed to percipients) were quiet, intellectual non-nervous people. With one particular sender the percipient guessed principally the "target card," with another the "card + 1," with another the "card - 1," supposedly depending on the psychoneural peculiarities of the sender.

The introduction of the concept of "precognitive telepathy" which, according to Soal, is sometimes replaced by "precognitive clairvoyance"[15] complicated the problem, and still further increased suspicions with respect to the very existence of mental suggestion. Guessing a card not yet pulled out of the pack! This involves foreseeing an as yet non-existent fact, one that could not have been predicted by any preliminary observation or reasoning! It was probably chiefly this which led G. Price to start talking about "the supernatural"—about a parapsychological "miracle."

However, is it necessarily "a miracle"? One parapsychologist, Raphael Kherumian, who is a materialist, one of the Members of the Council of the Paris Institut Métapsychique, tries to explain the phenomenon by the following train of reasoning:

The analogy between the telepathic sender and a radio transmitter of electromagnetic waves, and the percipient as a receiver, is inaccurate because in parapsychology "reception" sometimes forestalls "transmission," whereas there is no apparatus which could receive communications before they are sent. One must admit, so Kherumian writes, that it is difficult to refute this argument against the electromagnetic hypothesis of telepathy. But this difficulty may be overcome by moving in time the moment of transmission to account for "precognitive telepathy." Let us suppose that "preconscious" energetic processes which determine the future thoughts of the sender (which a moment later he begins to transmit to the percipient) can similarly act at a distance, just as they play a part in his present and past thoughts. Then, in cases of precognition, the brain of the sender would be something more like a tuning fork by which the percipient's thoughts are tuned.

According to Kherumian this postulate—according to which the stimulus giving rise to any parapsychological perception (including precognition) is not the result of psychical processes but must be attributed rather to their underlying energetic processes—makes paranormal cognition far more readily understood since they now cease to be strictly speaking "extrasensory." Precognitive ability now becomes comparable to sight or hearing and is "simply the quickest of the senses": it should not be forgotten that our various ordinary senses, in addition to possessing their own specific characteristics, also possess the peculiarity that one of them is "precognitive" as regards the others: sight is precognitive with respect to touch, warning us of the approach of, say, a person who begins to exist for the sense of touch only much later, and then only on condition that the approaching person actually touches us.[16]

I think it is necessary, however, to point out that this clever attempt of Kherumian's to understand parapsychological phenomena from an energetic point of view could have relevance only to a few of the instances where the sender himself is able to select the subject of his mental suggestion; it is true that in such instances it might be the case that "preconscious" mental energetic processes related to the transmitted subject arise in the sender and cause an actual image in the percipient more quickly than in the sender himself. But this explanation is not applicable to the very case which Kherumian is trying to explain—the Soal-Goldney experiments. In that case it only transfers the mystery from the percipient to the sender; the sender, not yet having turned up the following card and not yet knowing it, cannot have any mental processes caused by it—neither "preconscious" nor conscious. He has nothing to transmit to the percipient.

From what has already been said in this Chapter it will be apparent that techniques for establishing the reality of mental suggestion have undergone substantial changes with the advance of time. In the first period (the 1880s and '90s) attempts were primarily directed towards a study of spontaneous telepathic phenomena. But these are observed comparatively rarely, usually as a result of a severe nervous shock, something rather like a "mental storm," and it is impossible to re-create such a "storm" under laboratory conditions. From studying "macrotelepathic" phenomena observers turned to a study of "microtelepathic" phenomena, capable of repetition under specified conditions. However, experimental technique also has undergone substantial changes in the course of time.

For a long time qualitative methods of research prevailed, the results of which are hardly susceptible to statistical treatment. These are the methods of reproducing assignments that are being suggested mentally, such as pictures, objects, movements, acts of behaviour, etc. In such experiments the number of possible tasks is not limited by the instructions, but can be indefinitely large. The percipient is not told in advance what is the nature of the task, or only in the most general way. For example, an instruction may be given "Try to draw the object which I have in mind and which I will mentally suggest to you." Thousands of such experiments were carried out by many serious researchers, but they did not give, and actually could not give, a rigorous demonstration of the fact of the existence of mental suggestion.

Qualitative methods have a different purpose: they provide illustrations of the mental mechanisms involved in telepathic induction and perception. That is why they still retain their value to the present day.

However, to-day it is the quantitative methods of mental suggestion that have acquired the greatest importance and are most widely used. The quantitative results obtained in these experiments can conveniently be subjected to various statistical techniques, and were first introduced into parapsychology by Charles Richet in 1884. In quantitative work the percipient is not required to reproduce an object that is being transmitted by mental suggestion, but is merely required to identify one out of a limited number of objectives previously known to him. For example, the assignment might be to guess one out of the five symbols on Zener cards; one out of the five animals shown on Soars cards, etc. Experiments nowadays must conform to stringent conditions, otherwise statistical calculations may fail to carry conviction.

1. It is imperative completely to eliminate the possibility of any clues that might be perceived by the sense organs: unconscious whispering, the reflection of the object to be transmitted in spectacle lenses or the cornea of the experimenter's eyes, any mime or gesture, even sub-threshold stimuli that might lead to a correct answer. This is why the experimenter is separated from the experimental subject by a non-transparent screen or is placed in a cubicle or, still better, transferred to another room. The cards must be placed in non-transparent envelopes so that the subject should not even see their reverse side.

2. It is also necessary to eliminate the possibility of ordinary guessing by the percipient. For example, the subject must not be told during the experiment whether he has guessed the objects on the cards correctly or incorrectly: this

knowledge would enable the percipient to calculate which cards still remain in the pack and which have been taken out. (The pack usually contains in all 25 cards, i.e. each of the five figures is repeated five times.) A rule was therefore introduced that subjects must only be informed of the results after the termination of the experiment.

3. Probability theory is only meaningful when the targets are in random order. Consequently the greatest care is exercised in shuffling the cards. In the early experiments of Rhine and Soal the cards were shuffled by hand several times; subsequently they were shuffled automatically by a mechanical card shuffler. The method now in use is that, prior to the experiment, a packet of 25 cards is prepared by a person who is not participating in the experiment, by ordering the cards in conformity with printed tables of random numbers.

4. Moreover, in evaluating the results, other possible sources of error are taken into consideration, such as the observed preference of a percipient for a particular target of mental suggestion over other targets, as well as the random slips and errors generally, due to whatever cause, made by the experimenters when recording and checking the experimental results. In order to avoid errors of this type it is customary to prepare two reports of each experiment by two persons independently of one another.

In large series of experiments which complied with the above conditions American and English researchers have obtained above chance numbers of correct guesses of suggested objectives. For example, in a long series of guesses over a period of years, Soal obtained positive results the probability (or rather improbability) of which was $P = 10^{-35}$. Recently (1956–7) the same author obtained, in telepathic experiments on young cousins, (one of whom was the sender, the other the percipient), after carrying out 15,000 separate tests, an average of almost 9 successes out of each packet of 25 cards instead of the 5 expected by probability theory. In these experiments all 25 cards of the pack were guessed correctly twice; 24 cards were correctly guessed 4 times; between 19 and 23 cards were correctly guessed 40 times. These results, of course, enormously exceed what would be expected on probability theory.

Such marked above chance results are obtained in experiments using subjects who are specially telepathically gifted, but such subjects are rarely found. In experiments on ordinary subjects the results are lower (an average of 6 or 7 guesses to a packet of 25 cards); however, with a sufficiently large number of tests even such a result can be proved statistically not to be a chance result. In an equally large number of tests but with complete elimination of all telepathic and telæsthetic factors, the results are on the average no greater than 5 coincidences per 25 cards.

The value of the quantitative methods of Rhine and of Soal consists in the fact that they make it possible to detect very slight manifestations of telepathic ability in persons who produce nothing when qualitative methods are used.

In our own researches of the 1920s and '30s we also gradually passed from methods of reproduction to methods of identification, from qualitative to quantitative methods, from complex to simple techniques. In this respect we went even further than foreign researchers: they stopped at five targets previously known to the subject (Zener and Soal cards), whereas we reduced the number of targets to 2 (Mitkevich method: "white or black?"—hypnogenic method: "sleep!" or "wake up!"). Such methods of mental suggestion (selection from a choice of two assignments) can be called *alternative—one-or-other—yes-no*. They make it possible to collect very rapidly any amount of quantitative material suitable for statistical assessment.

Opinions concerning mental suggestion have undergone changes in conformity with the above-mentioned changes in methods of investigation. Contemporary researchers regard mental suggestion as one of the manifestations of psychoneural activity and that is all. They do not exaggerate its significance and do not see in it anything mysterious or mystical. This was, however, not so during the first period of studies. Barrett, Myers, William James, Lodge and many others saw in it a support for their idealistic and even religious opinions. They believed that the souls of the dead ("spirits") can telepathically influence the living in order to establish their "personality," their continued existence after death.[17] In the eyes of many sober-minded scientists this mystical romanticism compromised telepathic research for many decades. But this romantic shell has now fallen away and has given way to prose—to monotonous experiments repeated thousands of times with the scrupulous segregation of a few grains of telepathic "wheat" from tons of experimental "chaff."

Modern parapsychologists emphasize the very high variability and unreliability of the phenomena of mental suggestion. Percipients and senders who have given excellent results, far exceeding what would have been expected on probability theory alone, often give nil results in another experimental series, apparently carried out under exactly the same conditions—that is, results not exceeding what would be expected by chance alone. It has not yet been found possible to pin down the phenomena of mental suggestion in spite of many different attempts by a large number of researchers. It is now established that the success of experiments is affected by a great many factors of a primarily psychological nature.

It cannot be denied that there are individuals who are better than others at effecting mental suggestion (telepathic senders) and individuals who are better than others at perceiving mental suggestions (telepathic percipients). It is also hard to deny the existence of successful combinations of good senders with good percipients (so-called telepathic pairs).[18] The following is one of Soal's experiments supporting this contention. In this experiment several senders but only one percipient took part. Each sender in this long-drawn-out series received a previously shuffled pack of 25 cards in which five figures were repeated five times. The senders simultaneously turned up one card after another and simultaneously each one mentally suggested to the percipient the figure on the card he had drawn. In this experiment only one sender obtained an outstanding result far exceeding what would have been expected from probability theory alone. The other senders did not achieve results higher than chance, although they obtained high results with other percipients.

Nowadays great importance is attached to finding good senders and percipients, and to selecting effective telepathic pairs. In the U.S.A. there are even special offices engaged in this pursuit. A large-scale investigation of outstanding senders and percipients: is conducted by means of psychodiagnostic tests and various physiological and medical methods.[19] There has been a study of the effects of pharmacological substances which either stimulate or depress human psycho-neural activity on telepathic ability, and it has been established that the former increase and the latter decrease these abilities. It has been shown that an interest of sender and percipient in the experiments noticeably improves the results; when, after a time, the experiments begin to bore the participants the results decline sharply. Even such factors as the moods of the participants in the experiment and their relation with the experimenter and others present during the experiment, factors which can hardly be regulated, may affect the results: the presence of strangers, and persons who are sceptically inclined or who made fun of the experiments is particularly deleterious.

This is quite natural. Students of hypnosis have, after all, known for a long time that all the factors enumerated above affect the results of ordinary verbal suggestion made by a hypnotist to a waking subject. It is quite futile for sceptics such as Dr. Price and others to talk ironically about the fact that mental suggestion experiments are apt to fail in the presence of sceptical and hostile observers. They ask why do not severe examiners interfere with students taking an examination, whereas sceptics do interfere with the percipient in demonstrating telepathic abilities? In my opinion this is so because the mental processes leading to correct answers by examination candidates are subject to conscious mind and will, whereas the processes which effect telepathic perception are not so subject; they are subconscious and not subject to efforts of will. This can hardly be disputed.[20]

According to the data obtained from our experiments, far greater stability and regularity can be obtained by means of the hypnogenic method—mental suggestion of sleeping and awakening. This is quite to be expected: a subject in a hypnotic state is less susceptible than one who is awake to various outside physical and mental influences which interfere with the efficacy of verbal, as well as mental, suggestion.

It is to be regretted that the hypnogenic method of mental suggestion is no longer used abroad. There can be no doubt that its elaboration and elucidation is a great achievement of Soviet researchers.

Hypnosis is now more frequently employed for another purpose, and that is to improve the results of experiments carried out by the application of motor and sensory methods of mental suggestion. There are many indications to suggest that a light hypnotic sleep makes it easier for a percipient to receive telepathic impressions. Thus, for example, the Finnish psychotherapist Jarl Fabler found that "individual flashes of high scoring in guessing playing cards tend to occur in the hypnotic state"; out of 6 most pronounced deviations from chance expectations 5 were obtained by him under hypnotic conditions. The Czechoslovak parapsychologist Milan Ryzl employs hypnosis for the training and improvement of his subject's parapsychic abilities, including telepathic perception. It is suggested to hypnotised subjects that they will be eager to participate in the experiments and that they believe they will be successful, that they will acquire the ability to experience hallucinatory images corresponding to the suggested assignment, etc.[21]

Telepathic ability, from its very nature, is highly variable; it fluctuates periodically even during the course of one experiment. This was clearly manifested in the Soal-Goldney experiments to which reference has already been made, in which mental suggestions of pictures from cards were rhythmically transmitted at short time-intervals. It was found that the percipient, Basil Shackleton, whom we have mentioned earlier, successfully guessed cards when the sender selected them out of the pack at intervals of 2.6 seconds. When, however, the sender was instructed to slow down the rhythm to about 5 seconds per card, complete failure resulted. Moreover the percipient felt so irritated by the slowness of the new rhythm that he eventually refused to continue the experiment under these conditions. These observations enabled the authors to reach the conclusion that the success of experiments in rhythmic perception of mental suggestion sent rhythmically depends on the interrelation of two rhythms: the rhythm of the sender in sending suggestions, and the rhythms of suitable moments of perception in the percipient.

This does not mean, however, that every sender transmitting mental suggestions with the optimum rhythm is likely to obtain high better-than-chance scores. It is also necessary that the given rhythm of telepathic sending should correspond to the physiological rhythms inherent in the sender's organism. Thus, at the present time, "telepathic harmony" (l'accord télépathique) is understood as a special kind of resonance of physiological rhythms of sender and percipient (Warcollier, Kherumian) which is necessary for experimental success.[22]

These latter two French authors endeavour to explain the frequent failures of experiments conducted by means of the qualitative methods of mental suggestion (particularly the telepathic transmission of drawings), by a lack of coordination of the rhythms of sender and percipient.[23] The percipient's drawings only rarely reproduce the originals suggested; usually the reproductions are more or less distorted and do not represent the originals in a straightforward way but indirectly, by allusion, symbol, metaphor etc. We have already dwelt on this question in Chapter 5. Attempts to determine the psychological factors distorting telepathic perception and induction are continuing up to the present. As in the past, considerable attention is given to Freudian factors in explaining the results of telepathic experiments. Thus, at a symposium devoted to an assessment of the connection between psychology and parapsychology (Royaumont 1956) a whole series of papers dealt with this question: "Telepathy and psychoanalysis," "a psychoanalytic approach to *psi* phenomena," "the role of sexual repression in paranormal manifestation," etc.[24]

In order to eliminate as far as possible from telepathic experiments the various distortions and mutilations introduced by the consciousness of the percipient by his unconscious motives, by sexual symbolism, etc., experiments in mental suggestion abroad are nowadays often carried out on children of various ages,[25] on mentally and physically retarded adults and even on complete idiots, and frequently such experiments yield most clear-cut and significant results. I will mention only one example: Mme. de Thy, a University teacher of philosophy, describes the manifestations of astonishing telepathic perceptions by her mentally deficient brother,[26] whose mentality at the age of 47 was that of an 18 months old child. He is not capable of connected speech, and is only able haltingly to pronounce separate words. This, however, did not, according to his sister, prevent him from being a good telepathic percipient. 24 cases are cited in the article referred to in which he pronounced, with extraordinary speed and accuracy and completely without distortion, words and scientific terms completely unknown to him at moments when these words or terms happened to come into the minds of persons present. The author explains the absence of distortion of the telepatheme, so often displayed by other percipients, by the supposition that her brother's barely developed intellect was unable to control and alter the perceptions telepathically received. The instances given in her article occurred by way of spontaneous telepathy; no experiments in mental suggestion were ever deliberately conducted by anyone with this unfortunate individual.

I have not discovered in the foreign literature available to me any new researches in mental suggestion of voluntary movement of the type of the Joire or the Brugmans experiments; but modern parapsychologists again and again go back to a description of Bekhterev's old experiments in "mental" suggestion of behavioural acts on Durov's trained dogs,[27] without adding anything fresh.

There is a continuous increase in the number of investigations into mental suggestion involving the application of various laboratory methods of registration of physiological functions which do not appear in consciousness, and are not under voluntary control. Long ago in Holland the first attempt was made to obtain an electro-encephalogram at the time of occurrence of parapsychological phenomena.[28] In those days (1939) this method was still in its initial stage of development and could not contribute a great deal. The authors only succeeded in showing, apparently for the first time, that the electro-encephalogram obtained from supposedly telepathically gifted persons differed sharply during normal sleep, hypnosis and so-called trance on the one hand, and normal wakefulness on the other. The changes consist in the appearance of waves of great length—low-frequency—now known as the delta rhythm (see Chapter 2).[29]

I do not know whether these meagre results have ever been enriched by subsequent research. Some authors hope for results from experiments in which the electro-encephalogram of sender and percipient during the process of mental suggestion are simultaneously recorded, successful and unsuccessful instances being compared. It is expected that in successful experiments there will be synchronisation of the two rhythmic patterns. However, taking into account the limited scope of this technique, the complex nature of electro-encephalograms, the difficulty of explaining them from a physiological point of view and so forth, one cannot really expect a great deal. Perhaps more recently developed methods of electroencephaloscopy will be more promising since they make it possible to record the dynamics of electric potential simultaneously derived from a hundred or more points in the brain. Such a method holds out some promise of giving an accurate picture of the spread of bio-potentials in various functional conditions of the cortical hemispheres, which the electro-encephalogram cannot give.[30]

The technique of recording the galvanic skin reflex applied in experiments of mental suggestion by Brugmans et al. (1921), and later by us (1934),[31] was again made use of for a similar purpose by Woodruff and Dale.[32] Great importance should be attached to the results recently obtained by the Czechoslovak physiologist Stephan Figar[33] who simultaneously obtained on one and the same kymograph the plethysmograms (recording of changes of blood pressure in the blood vessels of the arm) of two experimental subjects. They sat back to back to each other at a distance of a few metres. In the course of his experiments Figar gave to one of his subjects (the sender) a note containing a task involving numbers, e.g. mentally multiplying two figure numbers. The mental effort brought about a lowering of the plethysmogram (i.e. withdrawal of the blood from the extremities due to a contraction of the blood vessels). This is normal, but what is extraordinary is that there was a similar lowering of the plethysmogram, after some delay, in the other subject, the percipient, who did no arithmetic and who did not know that such work was being performed by the other subject. This was observed in 33 percent of the experiments carried out, but with some pairs of subjects it happened considerably more frequently than with other pairs. A simultaneous lowering of the two plethysmograms was often noticed even without any mental arithmetic being done by either subject. If a so-called spontaneous plethysmographic reaction took place in one of the subjects (a lowering of the curve without ostensible reason) a similar reaction immediately began to set in with the other experimental subject. Such parallel shifts of the plethysmogram were observed in 185 instances, but in 106 cases there was no concomitant parallelism. At the request of the London Society for Psychical Research the experimental material compiled by Dr. Figar was studied in detail by Dr. D. J. West, one of the Society's Members[34] who approved of the experimental design and found the results convincing.

It should, however, be remarked that experiments of this nature, however interesting and however important, have but little to do with what is generally called telepathic connection. Telepathic connection is not only the energetic influencing at a distance of one organism by another, it also has a special informational character for living beings, or at least for one of a pair of them: one is informing another of some experience he is perceiving, or of a sensation he has received, of an image vividly experienced, of a feeling, a wish, etc.

The informational character is more clearly apparent in instances of spontaneous telepathy; in these circumstances the percipient fully or partially experiences, jointly with the sender, what is happening to the latter. The informational element is also clearly present, as demonstrated by the fact that the subject matter of mental suggestions may be carried out by various sensory methods, particularly the direct transmission of drawings, visual images of objects, or words or numbers. Even in experiments by means of motor and hypnogenic methods of mental suggestion the informational element is still present, as demonstrated by the fact that a subject often recognises who is the actual person who sent him to sleep, or compelled him to wake up at a certain moment. But in the experiments of Figar, and in similar ones, the informational element is reduced to nil: in such cases the percipient does not perceive the influence that has been exerted on him by the sender: in these experiments the influence itself contains no cognitional content and cannot therefore be called information.

The French parapsychological school, as represented by R. Warcollier and R. Kherumian, has been engaged for a long time in juxtaposing mental suggestion with other means of transmitting information at a distance, such as those used in technology. These authors have pointed out a number of analogies that exist between the activities of a telepathic percipient and the behaviour of cybernetic mechanisms. They believe that it is possible to construct a cybernetic artifact which would reproduce all the peculiarities, not only of ordinary (synchronous) telepathy, but also "precognitive" telepathy, together with all the defects to which the phenomena are subject, namely fragmentation, inversion and symbolisation of the information transmitted at a distance. In Kherumian's opinion, a percipient's susceptibility to "precognitive" telepathy has no creative aspect: there is in this susceptibility something rather more in the nature of a cybernetic calculating machine used by the scientist, than of the scientist himself using such a machine.

The French authors here cited attribute a special significance to an analysis of the defects and distortions characteristic of telepathic transmission, and

consider the so-called "negative deviations" (*l'écart négatif, psi*-missing)[35] established in Rhine's laboratory to be of special interest. This is the name given to
the strange fact that some percipients, in the course of quantitative telepathic
experiments, produce a percentage of failures considerably greater than would
be expected by chance alone. This phenomenon has not yet been explained
from a psychological, or a parapsychological, point of view. "And yet,"
Kherumian points out, "similar effects are now often achieved in technology:
we are thinking of missiles which can both follow and destroy a moving target,
or effectively avoid it."[36]

Kherumian still further emphasises the theoretical significance of "telepathic
defects." He says that spiritists are at a greater disadvantage when it comes to
explaining these than materialists. Why should a spirit, freed in the case of telepathy from the bondage of time and space, play about and make mistakes? This is
quite incomprehensible. But if one looks on telepathic abilities as no more than a
special organic sense comparable to our other senses, then the errors and failures
in telepathic experiments cease to surprise us. The sense organs known to us are
also subject to various shortcomings as regards their performances.

Attempts to apply information theory to the phenomena of mental suggestion and, still further, to construct a cybernetic model of telepathic
processes have already found some practical expression. Thus, Dr. Peter
Castruccio, one of the acting scientific collaborators of the Westinghouse
Corporation, U.S.A., created an electronic machine for statistical work on
experimental data of extrasensory perception, including mental suggestion. He
constructed a theoretical model of such phenomena by analogy with radar, and
applied modern communication theory in order to determine the working
mechanism of E.S.P. or "extrasensory perception."

Castruccio writes: "The analogy between electromagnetic waves as means
of communication and extrasensory perception is a purely formal one. It is
intended as expository, and not as depicting the essential nature of the phenomena, but merely their outward *modus operandi*. Science to date has shown
that those who follow a formal line of research obtain far more fruitful results
than those who attempt to uncover the essence of some phenomena. If we
learn how to apply E.S.P. we shall have made a great step forward, and we shall
not, in addition, have to pose the question how it happens in reality" [a good
example of the pragmatic practicality to which American thought is so
liable!—L. L. V].[37]

An unprejudiced reader will have noticed how very far the contemporary
approach to mental suggestion has travelled from its origins. We read in
Kherumian's article, to which reference is made earlier: "Parapsychology was
born out of beyond-the-grave questions. From its very inception, and up to the
present day, it has enjoyed the dangerous privilege of attracting almost exclu-
sively those who had hoped to find in this field proofs of their antimaterialistic
convictions" (p. 4). The same fully applies to telepathy which, in the days of

Frederic Myers and William James, had an intimate association with Spiritualism. What about the present day? Nowadays mental suggestion is not only exhaustively studied by means of experimental laboratory methods and statistical analysis, it also finds support in the most recent branch of scientific thought—information theory and cybernetics.

The question of the effect of the distance between sender and percipient on the success of experiments in mental suggestion has become a most lively topic of controversy.[38]

Modern quantitative experiments carried out in accordance with Rhine's method, as well as earlier qualitative methods of mental suggestion, seem to indicate that an increase in distance does not noticeably reduce positive results. For instance, in the experiments of J. G. Pratt, the percipient H. Pearce, whilst in other buildings at distances of 100 to 250 metres from the sender, successfully guessed the standard Zener cards. In 1950 tests were conducted in which a considerable "positive deviation" was obtained (188 more correct guesses than would be expected by chance alone); the probability of such results occurring by chance is insignificant.[39]

The idea that distance has little importance fully agrees with our own experimental data using the hypnogenic method (see Chapter 7).

Taking into consideration similar experimental data a modern parapsychologist, the late W. Carington, followed Barrett (see Chapter 8), and came to the conclusion that the "inverse square law" does not apply to these phenomena, and that consequently telepathic communication is not an energetic phenomenon, but is of quite a different nature. Carington thus contradicts not only the electromagnetic theory in all its forms, but all other physical theories of telepathic transmission. By declaring that suggestion at a distance is a non-spatial phenomenon he opens the door anew to idealistic notions of telepathy which separate spirit from substance and psyche from brain.[40]

The mathematician Hoffman,[41] in his article, challenges the Barrett-Carington argument. In his view their arguments are defective in that they confuse such different concepts as "intensity" and "intelligibility." Hoffman elucidates this by means of two examples as follows:

There is no doubt that the intensity of light is reduced as the square of the distance of the source of light. If the light energy is supposed to do some work, for example to dissolve the silver bromide of a photographic plate, it will accomplish such work the quicker and the more completely the nearer the photograph is to the source of the light: this is incontrovertible. But supposing the light serves merely as a signal, that it carries information about some event: in that case the length of the distance no longer plays an important part. We can understand the agreed meaning of a light signal to the same extent whether the signal is a blinding flash or a flicker so faint that we can only just perceive it. Signals, as information, are also subject to the inverse square law, but their intensity can be negligibly small.

Hoffman published his article eight years before the appearance of Norbert Wiener's famous book on cybernetics. We now know of the existence of energetic as well as cybernetic systems, and we know that in the living organism these systems are combined into one. "However, the specific characteristics and laws of these two realms of natural phenomena are different, and this difference must be clearly envisaged. The concept of information was developed later than that of energy, and the laws obeyed by cybernetic systems are far from adequately understood. So far only the foundations have been laid for such an understanding."[42]

Going back now to a question discussed earlier, we can assume that "a telepathic pair" at the moment of transmission of mental suggestion at a distance represents a temporarily active cybernetic system, with all the consequences that are entailed by such a concept. In particular, one can say that it is not thoughts that are transmitted—these are inseparable from the brain—but the information concerning thought, energetic signals.

In his second example the mathematician Hoffman approaches still closer to the question which is of particular interest to us. Let us imagine two radio transmission stations of equal strength: one nearby at A, the other at B at some distance. According to the inverse square law the signals from station B must be much weaker than those from A. But nowadays stations are provided with devices known as "automatic volume controls" which automatically intensify the signals weakened by distance from station B, thus rendering them equal in intensity to these from the nearby station A.

Such a set-up will "mask" the effects of the inverse square law, but obviously it does not cancel it. Hoffman continues his reflections by saying that it is permissible to suppose that a human being's physiological receptors of telepathic impulses contain something analogous to the design for an "automatic volume control," thus masking the effect of distance. Let us add to this that the role of such a set-up in the organism could easily be played by the well-known physiological mechanism which evens out the physiological responses to weak and strong stimulation, the "all-or-nothing law."

One can, then, agree with the Barrett-Carington position to this extent that, in order to effect suggestion at a distance the length of the distance itself has no great significance; but one can accept it with the very substantial reservation that this is the case not because the suggestion is transmitted by some non-energetic factor, but because the organism has arrangements which mask the action of the inverse square law, and that what is suggested at a distance—the telepatheme—is transmitted in the same way as signals and bits of information, and hence subject to the laws governing cybernetic systems.

This is, however, not universally accepted, and seems incomprehensible to those of our scientists who are inclined towards mechanistic materialism, or who do not wish to take cognizance of the achievements of cybernetics, and abroad the supporters of the "psychic," not to say frankly spiritualistic, hypothesis which separates the psyche from the brain, are not yet extinct.[43]

It is also possible to point to further hypotheses of a, so to speak, intermediate nature which use a concept of "psychic energy" and attribute to this a physical meaning. The hypothesis advanced by the electrophysiologist Hans Berger (who was mentioned earlier, see Chapter 2), may serve as an example.

The problems we have raised in Chapter 9 remain controversial: does a percipient know that the task he has accomplished (e.g. the Zener card he has selected) is in fact the very task suggested to him by the sender? And can a percipient tell exactly who is the sender?

According to our data, and also those of S. I. Turligin, such a possibility exists, but this is denied by many contemporary writers. They emphasize that the absence of a reciprocal connection is a characteristic feature of mental connection. The sender, so it is claimed, neither knows nor feels whether the suggested assignment has been correctly perceived by the percipient, (e.g. the right Zener card) and the percipient does not know either whether the card he has selected is the card the sender has suggested to him, or who is the sender.

I believe that in most experiments this is so, but that it can be otherwise. Some telepathic senders maintain that when an experiment is successful they feel a special "feeling of success." For instance, the psychotherapist Dr. K. D. Kotkov (see Appendix F) describes his experiences when effecting mental suggestion to induce sleep as follows: "I strongly wished that the girl (the percipient) should fall asleep. Finally the wish turned into a certainty that she was now asleep, and I experienced an unusual ecstasy of triumph at my success. I noted the time and stopped the experiment."

Moreover, our best percipient, Fedorova (and also S. I. Turligin's subject T. N.), "felt" when the sender started "working," and said so without ever being wrong. Apparently the hypnogenic method of mental suggestion is favourable to the manifestation of two-way communication between sender and percipient. In experiments employing other methods of mental suggestion the two-way character of the telepathic process is for some unknown reason not manifested.

It can be assumed that human beings and some animals have mechanisms or systems for telepathic communication; one for sending and the other for receiving the telepatheme. With some individuals (senders) the "sending mechanism" works better; with others (percipients) it is the "receiving mechanism." When one of those mechanisms in the sender or the percipient strongly dominates over the other, mental suggestion acquires a one-way character; when these mechanisms are both more or less equally well developed a two-way character of mental suggestion is manifested between sender and percipient: the sender telepathically influences the percipient and at the same time in some way experiences a reciprocal telepathic influence emanating from the percipient. Such an assumption is no more than a hypothesis requiring experimental verification; such confirmation could be obtained if one could find subjects who combined the ability of a good sender with those of a good percipient.

Attention must also be drawn to one further basic disagreement between parapsychologists, and this is closely linked with the foregoing problem: is the telepatheme directly transmitted from brain to brain ("intercerebral induction") without the intermediary of some receptors or sense organs, known or as yet unknown? Or should one accept the existence of some sort of special receptors specifically adapted for receiving the energy which transmits the information?[44] So far as one can judge, English and American parapsychologists hold the former view ("mental radio"), whereas the French school of parapsychology inclines towards the teaching of its founder, Charles Richet, concerning so-called cryptæsthesia (concealed feeling). Kherumian, for example, considers extrasensory perception, including telepathy, to be due to a special kind of feeling or sensibility which outstrips in speed of functioning all the sense organs known to us. In this respect the following extracts from his article (which has been cited repeatedly) is characteristic:—

"A great revolution in parapsychology will inevitably take place when it is proved that a sense modality exists the manifestations of which are not taken into consideration and have remained unknown to science. Let us consider for one moment what chaos would reign if our conceptions of the universe were worked out by deaf scientists who consistently denied the existence of sound and hearing! How many unsolved problems, errors and gaps there would be in all spheres of knowledge, from physics to æsthetics and theory of knowledge! We dare to contend that the development of parapsychology will some time turn the whole of our civilisation into a new path."[45]

At the basis of such optimistic pronouncements on the part of parapsychologists there lies the idea, whether realised consciously or not, that extrasensory abilities, including telepathy, are a progressive evolutionary acquisition from the animals; that in human beings this ability is more developed than in their zoological ancestors and that, in the future evolution of the human race, it will develop more rapidly and extensively. But is this so?

Evolution has provided animals and humans with three sense organs for distant perception: we can communicate at a distance by means of vision and hearing, and animals can in addition do so by scent. What could telepathic communication, supposing it to exist, add to these? What biological effects could it have? For animals it would have approximately the same effects that radio transmission has for modern man, which by far exceeds natural communication by means of the distance receptors and speech. Biologically this would be completely justifiable.

And in fact there are indications that some animals do possess natural "biological radio communication." Some puzzling phenomena in the life of certain insects may serve as examples, phenomena which have been known for a long time (Lawrence Harl and others), and known as "the call by a female of males at a distance." The Soviet biologist I. A. Fabri has studied this phenomenon in night butterflies for a period of six years.[46] On summer evenings a non-fertilised female was placed in a wire cage out on a balcony of a bungalow standing in a wood 5 km distant from two large villages. In 30 minutes or less males began to fly to her. In three evenings 64 males of this butterfly, rare in our country, were caught. They were marked with paint and carried away 6 to 8 km from the house, and set free. Some of the males returned. Flying slowly, which is characteristic for this species of butterfly, they covered the distance in 40 to 45 minutes. To accomplish this the males had to select the shortest direct path and to work their wing muscles hard in order to reach the female so quickly. It would appear that the female called the male in some unknown manner.

The males felt the call in this distant region, sometimes when there was no wind at all, and sometimes when the wind blew in the opposite direction to the smell (or "current" to use the author's expression) emanating from the female. It was established that it was the antennæ of the male that were the sense organs for perceiving the call by the "calling sender." Males whose antennæ had been cut off did not perceive the call of the female, and did not fly towards her (this was established long ago by L. Harl and acknowledged by Fabri).

What is this "calling" factor? Out of two possibilities, namely smell and electromagnetic signals, one must give preference to the one capable of acting against the wind, i.e. electromagnetic waves. However, this question has not yet been resolved and needs further research.

Recently it was established abroad that the olfactory organs themselves radiate waves 8 to 14 km in length. It is supposed that the part of a microantenna is played by the olfactory antennæ which are found in the mucous membranes of the nose. "This physical theory of smell," writes Prof. Frolov, "is also supported by some experiments on animals: for example, it is known that male butterflies can detect the weak scent of the female at a distance of several kilometres. This is incompatible with the theory that postulates chemical particles flying in the air."[47]

If all this is so then, in the case of the insect we are considering, the biological radio communication is effected by electromagnetic oscillations of micronic length, and not by means of some special receptors, but rather by the known olfactory sense organs acting as generators and receptors of the vibrations mentioned above. This is certainly not an instance of intercerebral (brain-to-brain) communication, but of a peripheral mechanism. However, this does not in any way exclude the possibility that at higher stages of evolution there might arise intercerebral radio communication using other waves of the electromagnetic spectrum. A simple signal—a kind of unconditioned reflex stimulus—is then replaced by a conditioned reflex stimulus referring to the second signal system; in short, it is replaced by mentally perceived information transmitted at a distance.

Now if one agrees that such phenomena as the calling of a male by a female in insects, and the acts of behaviour in the experiments on Durov's dogs in response to "mental" suggestion are indeed primitive varieties of telepathy, then it should be admitted that some animals are more highly gifted in this respect than Rhine's and Soal's best subjects.

In butterflies the calling of a male by the female is an important act in the life cycle of the species and is conducive to its preservation. In human beings instances of spontaneous telepathy are occasionally in the nature of a call for help, but more often the nature of the imparted information concerns experiences, often important, though sometimes trivial, which have an intimate personal significance. Though this also can have a certain survival value, nevertheless for modern man as such it cannot play a biological role.

The facts adduced and their explanation create an impression that being "telepathically gifted" is not a progressive element in the evolutionary process, but rather an atavism—a primitive attribute preserved in man from his zoological ancestors and resuscitated in some nervous or mentally deficient persons.

We find that the best subjects for our experiments in mental suggestion are to be found among psychoneurotics and even among the mentally retarded. It is hard to dispute this, and the observations of Mme. de Thy, to which reference has been made, are a clear example. However, this problem cannot be considered fully solved: examples of spontaneous telepathy are to be found in the biographies of outstanding persons. In healthy children, boys and girls, a heightened ability is sometimes noticed as a transitory phase: like provisional organs this ability later vanishes without a trace.

Parapsychologists are nowhere near solving another, probably more important, problem—that of the physical nature of the telepathic process. What is the nature of the energy that arises in the brain of the sender? How does the energy transmit telepathy through space? And how is it re-transformed into a psychoneural process after reaching the percipient's brain (or, on another interpretation, when it reaches the supposed peripheral special receptor organs whose specific function it is to perceive the telepatheme)? Our hopes for the electromagnetic hypothesis of telepathic phenomena, as we pointed out in Chapter 2, did not materialise—Cazzamalli's experimental results were not confirmed; rather our experiments, in which sender and percipient were screened by metal, yielded results which refute, and do not support, the electromagnetic hypothesis.

It is true that it is necessary now to introduce a substantial correction into the deductions we made in the 'thirties on the basis of the experiments using screening chambers. It was not then known that electromagnetic waves of low frequency and correspondingly great lengths (beginning with several hundreds of metres and more) are not completely absorbed by the iron and lead of the 1 to 3 mm thickness of the chamber walls we used and that, in a highly attenuated form, they could still penetrate into, and emerge from, the chamber.[48] It follows that the transmission of the telepatheme by an electromagnetic field of low frequency is not completely eliminated by our screening experiments.

There are, however, additional considerations that contraindicate the transmission of the telepatheme by low frequency electromagnetic fields. The researches of B. I. Danilevsky, F. P. Petrov and others, to which reference has already been made in Chapter 2, have shown that the brain and nerves are well screened by the surrounding fibres and liquids (lymph, blood) which possess a greater electromagnetic conductivity than the nervous tissues themselves. Consequently the enveloping tissues play the part of additional pathways (short circuiting connections) for the alternating current which is induced by the field in the nervous pathways within the skull.

This becomes obvious from the following simple experiment. A neuro-muscular preparation made from the body of a frog, when isolated in air (or in paraffin oil) is strongly affected by the field when the ends of the nerves and muscles are connected by a lead or thread wetted with physiological solution. The nerve cells and their dendites in the depths of the skull at most increase their rate of firing under the influence of a low frequency field of considerable intensity.

A high frequency field or, in other words, electromagnetic waves of short length (up to and including one millimetre) act upon tissue in a thermopene-trating manner (heating through). They do not have a stimulating effect (as a consequence of the square root law of Nernst); but again, they can alter the functional condition of the nervous elements by overheating. Some findings indicate "specific" activity of such short waves apart from heating; this is man-ifested when the intensity of the microwaves is 10 or more times lower than the threshold of their heating activity, and consists in resonance phenomena which appear in the albumen molecules of the surface layer of the cells of the skin (waves of 1 to 3 cm do not penetrate into the skin deeper than 2 to 4 mm).

Our experiments with screening by metal completely refute the con-tention of Cazzamalli and his followers that metre and centimetre waves could be the physical carriers in mental suggestion: excluded also are waves of 1 mm length, carrying large energies but still less capable of penetrating metal screens.

To this it should be added that quite recently (1960) the American physi-cists Volkers and Candib have published a fine paper entitled "Detection and Analysis of High Frequency Signals from Muscular Tissues with Ultra-Low-Noise-Amplifiers."[49] It was shown that contracting human skeletal muscles generate electrical currents of great frequency (up to 150,000 c/sec) and very low intensity (of the order of several microvolts). These currents were not detected until recently for technical reasons. It is of interest that small muscles appear to generate currents of greater intensity, e.g. the muscles of the little fin-ger. The brain, according to the data of these writers, does not generate high frequency currents of this sort, it merely regulates the intensity of such currents in the muscles.

The authors point out that similar currents generate radio signals around working muscle, and suggest that this might throw some light on the physical nature of telepathic phenomena. It will be appreciated that this suggestion is without foundation because, in the first place, the brain does not generate such high frequency signals, and also because radio signals of this wavelength are completely screened by a metal chamber, whereas the telepatheme, according to our data, penetrates into such a chamber without hindrance. A high-frequency field generated by muscles could, at best, suggest an explanation of pseudo-telepathic phenomena such as those of Reutler and Alrutz which were mentioned in Chapter 4.

According to Kherumian's contention "the electromagnetic hypothesis, which usually gives rise to pictures of wireless telegraphy, with a sender as transmitter and a percipient as a receiver, has at the present time been abandoned by most parapsychologists."[50] It is significant that Hans Berger himself—the famous contemporary electrophysiologist and discoverer of the electro-encephalographic technique—has abandoned it; and who could be expected to be a more ardent champion of the electromagnetic theory? Berger, in a short book published in 1940[51] developed the hypothesis of "psychic energy" as a factor carrying the telepathic information. He became interested in the question after some cases of spontaneous telepathy played some part in his life, and he personally carried out many experiments in mental suggestion on two hundred subjects.

Like Prof. Arkadiev, Berger considered that the changes of electrical brain potential were too small to explain the transmission over great distances of telepathetic information. He attempted to show that electrical energy, created by the cells of the brain, turns into psychic energy which can be diffused to any distance and passes through any obstacle which it may find in its way. Berger visualises this process as a diffusion of waves similar to Hertzian waves but not identical with them. He sub-divides the telepathic process into three stages as follows: (a) electrical brain processes are transformed into "psychic energy," (b) this energy diffuses in space, (c) when it reaches the percipient's brain it again turns into electric energy which causes physiological processes and mental experiences connected with it, and corresponding to the experiences of the telepathic sender.

Thus, according to Berger, "psychic energy" is the carrier of telepathic information which spreads in a wave-like manner, and which arises from a transformation of bioelectric potentials in the sender's brain and re-transforms itself into bioelectric potentials in the percipient's brain. In order to support the idea that the neurons of the brain use electric energy to produce psychic activity Berger quotes two phenomena, both discovered by himself: first, that the alpha-rhythm is suppressed at every mental effort, and secondly an increase in bioelectric potential (appearance of the delta-rhythm) occurs after loss of consciousness resulting from anæsthesia or some other cause.

Nowadays Berger's contentions have lost their force. As has already been pointed out in Chapter 2, the phenomena to which he refers are now explained by statistical factors (great numbers of cell generators of electric potentials, simultaneous reduction or increase of these potentials in the cerebral cortex as recorded by an oscillograph, etc.). Consequently Berger's hypothesis can no longer rely on the support of the factual data which he cites.

The so-called "meta-ætherial" hypothesis is somewhat similar. This was first advanced by Frederic Myers and has been further developed by French parapsychologists—Warcollier, Kherumian and others. According to this hypothesis the universe contains, in addition to the æther (repudiated by modern physicists) an ætheric medium of different nature which manifests itself in paranormal phenomena.[52]

Brain activity is supposed to be capable of causing vibrations in the meta-ætheric medium, which are transmitted in a wave-like manner through space and are, under certain circumstances, perceived by cryptæsthetic organs of sense,[53] highly developed in so-called sensitives, for example good telepathic percipients. The meta-æther and the processes occurring in it are regarded by Kherumian as an unusual physical medium and unusual physical processes which cannot as yet be detected by physical recording instruments, for which reason one has to make use of a living detector—the percipient's brain. This parapsychological materialist proposes a programme for studying "meta-ætheric energy." Here are his words: "The energy which, as we think, carries the parapsychic perceptions [including perception of telepathic information—L. L. V.] is undoubtedly not one of the forms of energy known to us. But there are many proofs that this energy has many similar characteristics. It is therefore, in our opinion, natural to investigate it, taking as a starting point its less puzzling aspects. The best means of approaching this problem and determining what are its specifically parapsychological features which cannot be accounted for by us, would be to construct instruments similar to the ones to which we are accustomed in cybernetics, so as to simulate telepathic transmissions just as they occur in reality, i.e. with numerous failures and characteristic distortions."[54]

Actually a similar problem often arises for contemporary physicists: they come across phenomena which demand the recognition of new fields and as yet unknown elementary particles within the atom. Experimental researches are made, which are sometimes successful. Let us recall the discovery of the meson field which was neglected at first and only experimentally confirmed after 10 years. Of course everything that exists in the universe is not as yet understood. Now micro-fields are being discovered not exceeding the boundaries of the atom: could one not suppose that sooner or later a new macro-field will be discovered which will go beyond the boundaries of atoms and engulf the surrounding space?

Some outstanding foreign scientists are already heading in this direction of research. For instance, Pascual Jordan, the German physicist and Nobel Prize winner, and Dr. B. Hoffman, a former collaborator of Einstein, think that a gravitational field seems to have some similarity with the force which transmits telepathic information, in that both act at a distance and penetrate all obstacles.

These examples, though they do not as yet amount to anything, show that the question of the energetic nature of mental suggestion is not a futile problem: it is beginning to exercise the minds of outstanding representatives of the most advanced branch of modern science—physics. And this provides a guarantee that, in one way or another, sooner or later, the problem will be solved.

Appendices

Appendix A

From an article on the First All-Russian Congress of Psychoneurology which took place in Moscow in January, 1923 (*"Izvestia,"* 1923, 16th January, no. 10) on direct thought transmission

A most interesting question was raised at the meeting of the Psychological Section on January 15th concerning the possibility of direct thought transmission at a distance, a natural scientific explanation of which now becomes possible in conjunction with a theory advanced by Academician P. P. Lazarev which explains nervous processes in terms of ionisation. K. Sotonin, a contributor at the meeting, read a paper *"Concerning alleged direct thought transmission in the experiments of Academician V. M. Bekhterev."* Sotonin considers direct thought transmission to be impossible and considers that the available facts concerning transmission can be explained by barely perceptible signals: movements of the eyes or the body, possibly unconscious.

Academician Bekhterev, in reply, pointed out that there are a tremendous number of facts and observations concerning direct thought transmission, and that such a question cannot be dismissed simply because it seems impossible. It once seemed impossible to fly in the air and the same applies, in the realm of mental phenomena, to hypnosis. Nevertheless, hypnosis is now applied in medicine and people do fly in the air. The solution of the problem of direct thought transmission is, according to V. M. Bekhterev, one of the most vital questions of the 20th century. In experiments carried out by him thought transmission was tested, not only in people but also in dogs. The experiments were carried out in such a way that the possibility of concealed prompting was eliminated. For example, the experimenter was placed in a closed booth, or the objects which had to be found were put in another room. Similar experiments were carried out in Moscow on Durov's dogs, also most successfully. Later on Prof. Kozhevnikov[1] took part in the discussion, and he also affirmed the fact of direct transmission of thought. Prof. G. I. Chelpanov, while not denying the facts, disagreed with the explanation offered by P. P. Lazarev. In view of the great importance of the question, Prof. G. I. Chelpanov proposed that a special report devoted to the subject should be presented at the next meeting.

From the resolution of the Second All-Russian Congress of Psychoneurology, approved on January 10th, 1924, in Petrograd (from records in the archives)

The Section on Hypnosis, Suggestion and Psychotherapy of the Second All-Russian Congress of Psychoneurology, having heard the Report on Mental Suggestion, deems it necessary to adopt the following resolutions:—

Attaching great importance to the question of so-called mental suggestion, the Committee [elected for drawing up the resolutions—L. L. V.] considers desirable a study of phenomena of this nature on strictly scientific lines, as well as participation in the activities of the International Committee, and the foundation of a scientific research group with a suitably equipped laboratory attached to the Russian Committee.

Appendix B

Introduction by Professor V. P. Ossipov to the Report for the year 1933–1934 of the Special Laboratory for the Study of Mental Suggestion attached to the Leningrad Institute for Brain Research (from records in the archives)

The work carried out by a group of collaborators at the Institute for Brain Research named after Bekhterev on the subject *"Psychophysical researches into telepathic phenomena"* is very extensive; it required a great output of energy and hard work because of the complexity of the essence of the problem and because of the immense difficulties in solving it either in a positive *or* a negative manner.

Although a very long time has elapsed since the question of telepathy was first raised, this question cannot be considered solved since the material collected is by no means immune to objections. In order to resolve the question one requires the experimental material to be studied scientifically and to prove the existence of a lawful connection of the phenomena in respect to some established principle. A negative solution must be equally strictly supported.

The group of workers at this Institute have not by any means in their researches solved the problem in either direction; the convincingness of successful experimental results is lessened by a series of unsuccessful experiments, although it would appear that inferences from statistics point in the positive direction. Furthermore, it was discovered that, in order to explain the positive results in a lawful context, it would be necessary to admit the existence of a new type of an as yet unknown energetic radiation—which also is contrary to the positive results of the researchers.

On the other hand, it would not be proper to judge the statistical preponderance of positive data to be due to chance.

Should the question eventually be settled in the negative, the negative solution would have to be strictly established.

In studying the research material obtained, it must be stated that, absolutely and without question, the researches conducted at the Institute for Brain Research were carried out in an entirely scientific manner, and that a strict research procedure was worked out involving a duly critical approach. It is necessary, however, to do a great deal of further minute detailed work to obtain a further elucidation of the problem.

1934
V. P. Ossipov
Director of the Institute for Brain Research.

Appendix C

On the question of telepathic phenomena. Suggested by academician V. F. Mitkevich, consultant physicist (from records in the archives)

With regard to psychophysical researches on telepathic phenomena I consider the following:—

1. The numerous psychophysical experiments carried out to date involving putting a percipient into a hypnotic sleep and waking him up, both by mental command from a sender, provide an absolute reason *for being considered sufficiently reliable proof of the reality of telepathic phenomena*,[1] regardless even of the fact that in some isolated instances there was spontaneous falling asleep and awakening of the percipient.

2. In view of the possibility of *subconscious* "agreement" between sender and percipient (directly or through a third person) the simple method of research into telepathic phenomena, referred to in paragraph 1, cannot by any means be considered satisfactory when the aim of the experiments is to ascertain the nature of the physical agency determining the telepathic connection, particularly in experiments with a bearing on the question of screening, of directed activity, of the effect of distance, etc.

3. A *complex method* might be more reliable and expedient, and I can visualise its realisation after the following manner:

(a) The percpient is put into a hypnotic condition by mental order of the sender.

(b) The sender then sets in motion a primitive roulette type of wheel which selects (without the intervention of the sender's choice) one eventuality out of a set of equally probable eventualities (e.g. a black or else a white screen in front of the sender's eyes).

(c) On the basis of the indication of the roulette wheel the sender orders the percipient to guess the given event, e.g., to guess "black or white?"

(d) This process of telepathic transmission of some visual impression (e.g. "black" or "white") from the sender to the percipient must be repeated as many times in succession as is necessary (taking into account preliminary experiments and probability theory) to establish as reliable the fact of transmission to the percipient of the chance impression of the sender.

(e) In order to avoid interference, such as that which might be due to subconscious suggestion from the assistant who is with the percipient, the attention of the assistant should not be concentrated on either of the equally probable events; for instance in the case of "black or white?" it is expedient to place before the assistant's eyes a perpetually moving screen with alternating black and white portions on its surface.

24th May, 1934.

Appendix D

On early experiments by French doctors in hypnotising and awakening at a distance

So far as I know, the first experiments were carried out as long ago as 1869 by the French doctor Dusard. He frequently put to sleep and awakened a patient of his, a young woman somnambulist, at times when she did not expect it. Mental suggestion was effected at various times and always gave the desired results. A description of some of Dusard's experiments can be found in Dr. Ochorovicz's well known book.[1] Here is one of these, described by Dusard himself:—

"From January 1st I stopped all calls on my patient G. and severed all connections with her family. I had no news of my patient. However, one day, on January 12th, when I was calling on other patients at the other end of the town, at a distance of about 10 km from the patient in question, the thought occurred to me: I wonder whether, in spite of the distance, in spite of the fact that I have severed all communication and that a third person is now involved (the patient's father was now magnetising his daughter instead of myself), I could compel the patient to obey me as heretofore? I mentally forbade the patient to respond to the influence of her father in sending her to sleep; but within half an hour I realised that, just supposing my order was complied with, I should be harming the sick girl. So I decided to countermand the order and thought no more about it. How great was my surprise when a messenger came to my house next day with a letter from G.'s father. In the letter he wrote that on the previous day, the 12th of the month at 10 a.m., he only managed to send his daughter to sleep after a long and painful struggle. When the patient was finally asleep she had said that she resented the order and that she only fell asleep if I allowed it."

This experiment is unlike the others: there is here no sending to sleep or waking up, but a mental suggestion that it will be impossible to go to sleep— a suggestion of insomnia. In 1878 another French physician, Dr. Héricourt, a collaborator of Prof. Janet's, caused a hysterical 24 year old patient of his to fall into a hypnotic sleep by distant suggestion; he caused her by means of a mental order to leave her flat, to go out into the street and to go along the street to a certain place. Those early experiments did not attract sufficient attention, but they laid the foundation for the well-known experiments of Janet and Gibert, and later of Charles Richet, with the subject Léonie B. These experiments were mentioned earlier, in Chapter 6. In 1886 Frederic Myers, the most active member of the London Society for Psychical Research, and Julius Ochorovicz, the famous Polish psychical researcher and professor of psychology in the University of Lvov, came to Le Havre in order to take part in these experiments. Both published in the press descriptions of their observations. Prof. Ochorovicz writes about Léonie B. as follows[2]:—

"This subject is a simple peasant woman from Brittanny, about 50 years old, very healthy, honest, modest, not stupid although she has received no education, (she could not even write, though she could recognize some letters); she is strongly built; she had suffered from hysteria in her youth but had been cured; she has a husband and children, all of whom enjoy perfect health; she is very easily hypnotised: it is sufficient to hold up her hand and press it lightly with the suggestion of putting her to sleep. Within 2 to 5 minutes her vision becomes blurred, her eyelids begin to flutter rapidly until finally the eyeball is concealed underneath the lid; her chest begins to heave, she behaves as if she were falling ill; she sighs and falls backwards into a deep sleep" . . . . . . [There follows a description of one of the experiments —L. L. V.]

"I found Drs. Gibert and Janet, who were already definitely convinced of the undoubted effects at a distance, ready to accept all the conditions which I suggested to them, and they allowed me to verify the phenomena in any way I wished. Mr. F. Myers, Marillie of the Psychological Society, another physician (A. T. Myers) and I formed a committee, so to speak, and all details of the experiments were jointly established by us. The following are the safeguards which we took in our experiments:—

1. The hour for the action at a distance was determined by casting lots.

2. Gibert was told the time only a few minutes beforehand, and the members of the committee immediately went to the cottage where the subject lived.

3. Neither she nor any of the other inhabitants of the cottage, which was a kilometre distant from our house, knew either the time of the experiment or of what it would consist.

In order to avoid unintentional suggestion the members of the committee only entered the cottage for the purpose of establishing whether sleep had occurred.

We decided to repeat Héricourt's experiment to put the subject to sleep [by mental suggestion—L. L. V.] and to make her go through the whole town.

It was 8:30 p.m.; Gibert agreed to our conditions. The hour was determined by casting lots. Mental activity was to start at 8:55 and to continue until 9:10. At the time there was no one at the cottage but Léonie B. and a cook who was not expecting any activity on our part. Nobody entered the cottage. We approached the cottage at 9 p.m. All was silent. The street was deserted. Without making any noise whatever we separated into two groups in order to observe the house from a distance.

At 9:25 1 saw a shadow appearing at the garden gate: it was she. I hid behind the corner in order to be able to hear without being seen.

But there was nothing to hear: the somnambulist, after stopping for a while at the gate, went back into the garden. (At that exact moment Gibert stopped exerting his influence on her: as a result of the strain of thinking he fainted—or rather dozed off until 9:35.)

At 9:30 the somnambulist again appeared at the gate; this time she hurried out into the street without hesitating and with the speed of a person who is late and who must without fail attend to important business. The other members of the committee had no time to warn us—myself and Dr. Myers. But, having heard hurried footsteps, we followed the somnambulist who did not notice her surroundings, at any rate she did not recognise us.

When she reached Bar Street she began to sway, stopped and almost fell, then she suddenly resumed walking. It was 9:35 (at that moment Gibert had suddenly come to and begun to work on her again). The somnambulist walked quickly and paid no attention to anything.

In ten minutes' time we reached Gibert's house while he, thinking that the experiment had failed and being astonished that we had still not returned, came to meet us—and collided with the somnambulist who had her eyes closed as before.

She did not recognize him. Absorbed in her hypnotic monomania she hurried towards the stairs. We all followed her. Gibert wanted to enter his study but I took him by the arm and led him into another room.

The somnambulist, who was very agitated, looked for him everywhere, bumping into us. She feels nothing. She then enters the study and repeats in a sorrowful voice: 'Where is he? Where is Dr. Gibert?'

Meanwhile the hypnotist is sitting motionless, bending down. She enters the room, almost touching him in passing, but her agitated condition prevents her from recognising him. Then it occurs to Dr. Gibert to summon her towards himself: whether as a result of his effort of will or by sheer coincidence she returns and takes him by the arm.

She is seized with great joy. She clasps her hands as a child would do and exclaims: 'Here you are! At last! I am so pleased!'

I am convinced at last [adds Dr. Ochorovicz] of the wonderful phenomenon of psychical activity at a distance, which creates a revolution in our present day accepted opinions and ideas."

"The correlation between separate observations is for us the best criterion of truth" wrote the well-known psychologist A. Binet in his article *"Experimental Psychology at the London Congress of 1892."*[3]

In accordance with this wise remark, I quote a second observation of the very same experiment written by another participant: F. Myers.

"B., a well-known subject of Prof. Pierre Janet, was the subject in these experiments in telepathic hypnosis. The experiments were carried out in Le Havre by Prof. Janet and by Dr. Gibert who conducted the experiment.

I quote below extracts from my own experiments of April 20 to 24, 1886, carried out jointly with Dr. A. T. Myers which were written in the form of reports, and which form the basis of an article in the journal of the Society for Physiological Psychology, dated May 24th, also published in the Proceedings of the Society for Psychical Research, Vol. IV, pp. 131–137.

On the evening of April 22nd we all had dinner at Dr. Gibert's. After dinner he made his standard attempt to put B. to sleep at a distance from his house in Série Street and to call her to him by an effort of the will. The subject was at that time in her cottage in La Ferme Street. At 8:55 Gibert went to his study and Ochorowicz, Marillie, Janet and Myers went towards the cottage and waited in the street by the house. At 9:22 Dr. Myers spotted B. who appeared at the garden gate and then disappeared again. Those who observed her close by saw that she was obviously in a somnambulistic state. She wandered about the garden, murmuring something. At 9:25 she emerged (her eyes were closed all the time as far as one could see); she quickly passed Janet and Marillie without noticing them, and went to Gibert's house, not by the usual and shortest route. (It emerged later that the cook saw how she entered the sitting room at 8:45, left it at 9:15 and did not return again.) She avoided street lamps and the traffic, and crossed the street many times. Nobody went up to her and started a conversation. Within 8 or 10 minutes her walk grew less steady, she stopped and seemed about to fall. Myers noted the time on his watch: it was 9:35. At about 9:40 she grew bolder, and at 9:45 she reached the street where Dr. Gibert lives. There she met him but paid no heed to him and entered the house, where she began to run from room to room on the first floor. Gibert did not touch her hand until she recognised him. Then she quieted down. Gibert said that from 8:55 to 9:20 he had thought about her intensively, and from 9:20 to 9:35 he thought about her less strenuously. At 9:35 he stopped the experiment and started to play billiards, but after a few moments he started again to call the subject. It was later established that his presence in the billiard room coincided with the time when the subject had displayed uncertainty in the street, but this coincidence might have been fortuitous.

Out of a series of 25 similar experiments 19 were successful. Experiments were carried out at various times of day and at different intervals in order to avoid the possibility of the subject expecting the beginning of an experiment."[4]

A juxtaposition of the reports made by Ochorovicz and by Myers shows that these notes coincide as regards the main points and supplement each other with respect to detail. However, the times in minutes noted in the reports do not always correspond exactly; sometimes the discrepancy amounts to as much as five minutes. It is not mentioned whether the watches of Ochorovicz and Myers were synchronised before the experiments.

I will quote one more observation, made by Charles Richet in the hospital at Beaujon, apparently using another experimental subject.

One day, whilst sitting with friends at lunch in the hospital's dining room, (his colleague Landusi, at one time one of the doctors at Beaujon hospital, also being present) Richet declared that he would undertake to put one of the patients to sleep at a distance and make her come to the dining room solely by an act of will. After a lapse of 10 minutes nobody came. The experiment was considered unsuccessful. However, in reality it turned out to be otherwise: a few minutes later someone came in to say that a sleeping patient was roaming the corridors and looking for Dr. Richet whom she could not find.[5]

In the experiments with Léonie B. the results obtained by Richet were poorer than those of Janet and Gibert (16 successes in all, out of 36 experiments, as against 19 successes out of 25 in the Janet and Gibert experiments). In Richet's experiments it took 11 minutes to put Léonie to sleep by mental suggestion. In Janet and Gibert's experiments the first signs that the subject was going to sleep were already noticed after 2 to 5 minutes, depending on whether it was Gibert or Janet who was putting her to sleep.[6]

In our experiments our subjects Fedorova and Ivanova usually fell asleep within 1 to 2 minutes (see Fig. 28), rather more rapidly if Tomashevsky was the sender than if it was Dubrovsky.

Appendix E

About telepathy. First public experiment demonstrating mental sending to sleep and awakening in the U.S.S.R. by Professor K. I. Platonov[1]

In connection with the debate on telepathy which took place in Kharkov [in February, 1961] some recollections have come to mind concerning an experiment which I conducted at the All-Russian Congress of Psychoneurologists, Psychologists and Teachers (in Leningrad in December, 1924).

At that Congress I confined myself to one demonstration without making any attempt to explain the essence of this mysterious phenomenon because this would have gone beyond our objectives at that time.

My experiment was carried out by a definite method, both before a large audience of the Hypnological Section of the Congress, and also at the laboratory at Leningrad University with the participation of the President of the Congress, V. M. Bekhterev, who became interested in the demonstration.[2]

Amongst those present were members of the Congress L. L. Vasiliev, E. S. Kotkov, V. A. Podierny and others.

The experiment was carried out in the lecture hall where the meetings of the Hypnological Section took place. Prof. A. V. Gerver sat at the presidential table facing the audience. The subject M. sat at the same table, half facing Prof. G. who was talking to her, and sideways towards the audience. Behind her back at a distance of about 6 m there was a blackboard, edge-on in relation to the spectators. I stood behind the blackboard, in full view of the audience, but outside the range of sight of the subject. It had been arranged with the audience, prior to the arrival of M., that when I silently covered my face with my hands, this would mean that I had begun the experiment of mentally sending her to sleep. Having covered my face I formed a mental image of the subject M. falling asleep while talking to Prof. G. I strenuously concentrated my attention on this for about one minute. The result was perfect: M. fell asleep within a few seconds. Awakening was effected in the same way. This was repeated several times.

During the laboratory experiments I, as the sender, was in a closed cabin, with an electric signalling lamp (a pocket torch). The switch by which this lamp was switched off was outside the chamber under the top of a table which stood next to the cabin. V. M. Bekhterev, who was signalling to me by switching on the lamp, sat with his back turned to the switch on which his fingers were resting (behind his back). The subject, M., who was supposed to perceive my mental suggestion, was sitting opposite to Bekhterev. The signals were given me by Bekhterev while he was talking to her, and such signals were not regular, i.e. they were given at irregular intervals of time.

When I had received a signal I closed my eyes and mentally represented to myself the image of the sleeping subject. According to the statement of those present she immediately stopped talking and fell asleep,

Awakening was effected in the same way, i.e. by my representing to myself an image of M. waking up from her sleep. This experiment was repeated three times with the same success. When questioned by those present, "What happened to you just now?" M. answered: "I don't know, I think I must have slept."—"Why did you fall asleep?"—"I just did. I felt like going to sleep."

It is important to note that when I tried to influence the subject by means of a mental command, such as "go to sleep—" or "sleep!" such orders invariably remained without result. But when I visualized the image of the sleeping M. (or of M. awake as the case might be) the effect was always positive.

As regards M.'s special personality traits, I ought to say that she was one of my patients and proved to be a good somnambulist and she was very suggestible. I had first met her in the previous year, in 1923, at a Slavonic resort where I was working as a consultant in nervous diseases, and there I demonstrated a number of the usual hypnotic suggestion effects on M. At that time I already knew about the method of "mental suggestion." This method was tried out by me on M., who quite unexpectedly gave positive effects in falling asleep simply as a result of my mentally representing to myself the visual image of M. sleeping. At that time I became convinced that this effect could not be obtained merely by means of a mental order to go to sleep, whereas when I employed the method of forming a mental representation of the desired end this proved to be perfectly effective.

It should be noted that the effect of waking up M. also occurred quite suddenly within a few seconds after I had started mentally visualising her awakening.

These experiments of mine were carried out with the collaboration of my son, K. K. Platonov, a biologist. The duration of the latent period that elapsed between the onset of my mentally visualising her as falling asleep (or awakening) and the moment of the occurrence of the corresponding effect, was eventually recorded, and proved to be equal to several seconds.

It should also be noted that M. fell asleep not only when she was quiet, but also when talking to someone. Once she was put to sleep in this way by me while she was waltzing to the tune of a piano. Furthermore, I did not have to be near her, but could be a considerable distance away from her, not even necessarily in the same house. All this surprised us tremendously as well as the other doctors in the resort.

It should be particularly stressed that M. was entirely unaware of the nature of the experiment. This was done without her knowledge and without any preliminary conversation on the subject. She had thus no idea of the meaning of the experiment, and she herself had no idea what caused her to fall asleep.

On my arrival at Leningrad for the Congress in December 1924, I quite accidentally met Miss M. in Leningrad, in which town she happened to live. She willingly agreed to go to the Congress with me and to be present at one of the meetings which would interest her. She was not told the purpose of my taking her to the Congress.

After the demonstration carried out at the Congress which was described above, she was puzzled and asked me "Why did you invite me to the Congress? I don't understand it! What happened to me? I slept, but I don't know why—you did not send me to sleep!"

Thus at the Congress we confined ourselves to but one demonstration of this incomprehensible and puzzling phenomenon, regarded as doubtful by many of the watching members of the Congress, and definitely accepted by only a few isolated individuals. But the phenomena observed in our experiments were so striking in their clarity and consistency that it was impossible not to admit their reality.

Later on, after the Congress, M. acted as experimental subject for L. L. Vasiliev, a physiologist, who had been present at our demonstration at the Congress. These demonstrations led L. L. Vasiliev to undertake a serious study of telepathic phenomena, manifested, as is known, in the form of thought transmission at a distance. The subsequent numerous researches in this field carried out by L. L. Vasiliev (1926 to 1960), with positive results, in mental sending to sleep and awakening at a distance, cause one to consider the phenomena extremely seriously. Recently L. L. Vasiliev has written to me that of all his numerous experiments in telepathy pride of place, as regards the number of successes, is taken by sending to sleep and awaken.ng.

As I see it, all these data may help to contribute towards an acknowledgment, not often to be found, of the reality of mental sending to sleep and awakening at a distance of subjects. They give us also, however, the right to search for means of finding a scientific, materialistic grounding, not only for the phenomena of telepathically inducing sleep, but for many other telepathic phenomena also.

Apparently at the present time conditions are favourable for a study of these up to now "mysterious" phenomena of telepathy. Until recently it was thought that the electrical processes occurring in the cerebral cortex were at the basis of this phenomenon. It was thought that the cortex produces various electric waves, but in recent times a supposition has arisen according to which "a physical field, new to science and produced by the brain" may act as a transmitter of telepathy.

However, it should be borne in mind that at the present time the solution of this most intricate problem of "thought transmission at a distance" is still in its initial stages.

I will permit myself to express some speculations concerning the phenomenon of sending to sleep and awakening at a distance which arises out of the method of distant sleep induction applied by us.

First of all, it should be noted that in our experiments the influence was exerted by means of visual images, i.e. with the participation of cerebral mechanisms connected with the visual analyser, whereas in the usual method of sending to sleep it is the mechanism connected with the vocal analyser that was called into operation.

If one admits visual telepathic dreams one is automatically struck by the similarity in the mechanism of formation by means of the brain's analyser. The mechanism for the occurrence of such dreams may likewise be regarded from the point of view of the science of electrical processes arising in the brain. Accordingly the source of such dreams could be found in the brain of another person who is experiencing some unusually strong emotion, and who stands in a close family or other intimate relation to the given person.

In conjunction with this, another thought may be expressed: the analogy between cerebral "tele-vision" and the mechanism of modern television communication. The screen of a television set may be taken as corresponding to the brain of the percipient in an experiment of tele-induction of sleep. All this is, of course, hypothetical, but it will perhaps be considered in due course. It is not true that a television set also reproduces reality on its screen by means of radio communication?

One cannot fail to notice that the instructions involved in mental suggestion require not only verbal suggestion but visual suggestion also. Could it not therefore be the case that some influence emanating from the visual area of the sender's brain lies at the root of this phenomenon, since the representation in me—as sender—was so strongly connected with the visual images of the subjects?

In view of this, there would seem to be good reason for supposing that, in experiments of tele-induction of sleep, the mechanism of suggestion is effected by means of the cortical area of the visual analyser, and not the auditory, motor or speech centres.

I would mention that some of those who spoke at the Kharkov discussion on telepathy apparently considered that the phenomena of mental suggestion are connected with the science of hypnosis and verbal suggestion. Prof. H. P. Tatarenko, who took part in the discussions, quite properly denied such a connection. Of course telepathic phenomena are essentially different from those of hypnosis and verbal suggestion.

Perhaps the above suggestion of the analogy with ordinary television may prove helpful in the matter: analogies are sometimes useful. At a certain stage in the struggle for scientific truth conclusions based on analogical reasoning are permissible.

Appendix F

Concerning experiments in mental suggestion of sleeping and awakening carried out by scientists at the University of Kharkov

(Letter from Prof. K. I. Platonov to L. L. Vasiliev)

Kharkov, 31st January, 1961.

Dear Leonid Leonidovich,

On my return from Leningrad in 1924, where I had demonstrated my experiment of thought transmission at a distance (primarily sleeping and awakening) at the Congress, similar experiments were carried out in Kharkov by one of my medical research students in the science of hypnosis and suggestion. I am sending you the information concerning these experiments, having received them from Konstantin Dimitrievich Kotkov, an active participant in the experiments. I am sorry that, as you will see for yourself, the material is only descriptive. Nevertheless it seems to me that this information may not be redundant. I was kept informed concerning these experiments, and took some part as consultant, the three experimenters (one a physicist in the faculty of natural science, the others physicians: an endocrinologist and a psychiatrist) having heard my paper and witnessed my demonstration in Leningrad in 1924.

I also had some notes on the Kharkov experiments with Mikhailova, but they have vanished without trace. You wrote to me that your greatest successes were achieved in connection with mental sending to sleep. Apparently there is a conformity to law.

I wanted to tell you all the above, my dear Leonid Leonidovich. I cordially shake your hand and embrace you. I enclose herewith the letter from Dr. K. D. Kotkov.

Yours
K. Platonov.

(Letter from Dr. K. D. Kotkov to Prof. K. I. Platonov)

Dear Konstantin Ivanovich,

In 1924 A. V. Dzhelikhovsky, professor of physics at Kharkov University, and Dr. L. P. Normark, a chemist, carried out a small but most interesting piece of work in thought transmission at a distance. The principal instigator of the work and its heart and soul was Normark. I was invited by them to participate as an experimenter. That is why they kept all the experimental reports. I am sorry to say it never occurred to me to make special copies for myself. Dzhelikhovsky died during the German occupation of Kharkov, and I learned of Normark's recent death from you when we last met. Therefore there is now no one from whom one could obtain the reports of this remarkable work: I do not know of their fate.

I know that my communication, unsupported as it is by precise documentation, is of no value as being mere hearsay, but I will give you my recollections. I shall be extremely pleased if they prove to be of any use to you for any purpose whatever.

A few words about the experimental subject. She was a student at the University, 18 or 19 years old, absolutely healthy, of varying mood, predominantly somewhat depressed, a perfect somnambulist. In Normark's laboratory I was able, in the presence of students, to give really marvellous demonstrations of experimental suggestion. Dzhelikhovsky and Normark asked me to be the experimenter in their work. I picked out this girl as my subject. But how to proceed after that?

We thought it impossible to experiment on her without her knowledge. We asked her permission, and she gladly agreed. She knew that we were going to experiment on her for two or three months, but where and when and on what days and at what times, and in just what the experiments would consist, she did not know.

She used to fall alseep instantly under the influence of mental suggestion transmitted to her, and she woke up instantly under the influence of mental suggestion. She remembered nothing. Here is an example: she once fell asleep in Dzhelikhovsky's flat while standing up and holding up a test tube containing some preparation which had been shown her by Dzhelikhovsky, and which she was examining with great interest. When she awoke she continued to examine it as though nothing had happened and imparted her impressions to the professor, continuing her conversation with him at the point where her thoughts were interrupted. She asked us several times: "Well, when are the experiments about which you gave me warning going to start?" We usually replied to this question: "We'll wait a little longer, the apparatus isn't ready yet." So from the very beginning of the experiment to the very last she did not know whether any experiments had been carried out with her, and what kinds of experiment they were. After that she left, and I lost contact with her.

A few words about my methods of thought transmission. I used to sit in a comfortable arm chair in complete silence. I closed my eyes. I mentally murmured to my subject the words of suggestion, "sleep, sleep, sleep!" This I will call the first factor of mental suggestion.

The second factor: I represented to myself the image of the subject with the most vivid hallucinatory or hypnagogic intensity. I pictured her to myself as being fast asleep with closed eyes.

Finally the third factor: I consider this the most important one. I will call it the factor of wishing. I strongly wished the girl would fall asleep. Finally this wish turned into a certainty that she was now asleep and experienced a sort of unusual ecstasy of triumph at my success, I noted the time and stopped the experiment. The time was accurately noted. I waited for the signal to proceed with the awakening of the subject, and she similarly woke up at the same moment as my signal. All these three factors acted simultaneously for a length of time from 3 to 5 minutes.

Not very many experiments were carried out—not more than 30. Not a single one was a failure. The intervals between them were between 1 and 3 days; sometimes we experimented on two days in succession.

We carried out experiments not only in sending the subject to sleep and waking her up, but also in summoning her. While in my own flat, I summoned the girl to Normark's laboratory at a time precisely agreed upon with Dzhelikhovsky and Normark. I used the same method. When the 'ecstasy of triumph at my success' occurred, I stopped the experiment and went to the laboratory. I usually found the girl already there, or else she arrived a little after my arrival. When she was asked why she had come, she generally answered, looking embarrassed: "I don't know. . . . I just did. . . . I wanted to come. . . ."

None of us doubted that thought transmission is possible. Our aims were to confirm such a possibility and, principally, to determine the essence of the energy which at that time radiates from the brain.

The subject used to be invited to the experiments under pretexts which could not suggest to her any reasons for supposing why she had been asked. At the actual times of the experiments the girl's attention was kept occupied, as much as possible, by anything which might interest her. She was given no opportunity of concentrating on anything of her own choosing.

The first experiments were carried out in the same building. We were separated by a few rooms. We then passed on to experiments in which we were at different ends of the town. The success was the same. Communication was so well established that we did not miss any possible observations. Everything was pre-arranged in a most precise manner.

The second part of the experiments was carried out with the use of apparatus. I was placed in a sort of cabin specifically prepared for the purpose, which insulated me from the rest of the world. Changes in the arrangements of the cabin were made when they seemed necessary. Mental suggestions were invariably successful with the same astonishing speed and precision.

Only one possible defect worried all three of us in this work: that the girl could have been warned of the experiments. Yet everything was done so skilfully and accurately that she could not have known of the experiments. This is established by the fact that she asked us up to the last moment, up to the very last experiment, when at last our experiments with her would start.

Normark and Dzhelikhovsky reached definite conclusions concerning the possibility of thought transmission at a distance without the intervention of the sense organs, directly from one brain to another by means of radiations from the brain at the time of mental effort, and the effect of such radiations upon the brain of the subject. They also reached definite conclusions about the nature of these radiations, but all this has remained in the research material—in the reports of the experiments and the authors' conclusions.

After these experiments, in that same year, Dzhehkhovsky presented a separate paper to a scientific conference of physicists which took place at Kharkov University. The work was intended for publication somewhere, but I do not know anything further as to its fate.

Here, I am sorry to say, is what little I can contribute. If you would like any further details I shall be glad to let you have them.

Yours faithfully,
(Dr.) K. Kotkov.

Notes

Notes to Preface

(1) V. M. Bekhterev (1920), Experiments on the effects of "mental" influence on the behaviour of dogs, *Problems in the Study and Training of Personality*, Petrograd, 2nd Edition, pp. 230–265.

(2) P. Flecksor (1920), Experiments in so-called mental suggestion on animals, *ibid.* p. 272; A. G. Ivanov-Smolensky (1920), Experiments in mental suggestion on animals, *ibid.* p. 266.

(3) Similar experiments were carried out at the same time in Moscow at the Laboratory of Applied Comparative Psychology, under the direction of V. L. Durov with the collaboration of B. B. Kajinsky, an engineer, who was one of the pioneers of the study of mental suggestion in the U.S.S.R. See: B. B. Kajinsky (1923) *Transmission of Thought*, Moscow.

(4) V. M. Bekhterev (1921), *General Reflexology*, Petrograd, p. 263.

(5) The text of this resolution is to be found in Appendix A, p. 141–142.

(6) Also called "parapsychology."

(7) See V. P. Ossipov, Introduction to the Report of the Laboratory for the year 1934 (Appendix B, pp. 143–144).

(8) Appendix C, pp. 145–146.

(9) This paper was subsequently (1936) presented by I. F. Tomashevsky as a thesis at the Institute for Brain Research.

(10) B. B. Kajinsky (1959), Wireless thought transmission, *Comsomol Truth*, Nov. 15; M. Guzeev (1960), Tomorrow's psychology, *Leningrad University*, June 15th; Interview with B. B. Kajinsky (1960) *Science and Life* No. 11, p. 46; Discussion: Thought transmission—is it possible? in *Knowledge is Power* (1960), No. 2., pp. 18–23; Interview with Prof. L. L. Vasiliev (1961), Do *psi* phenomena exist? *Shift*, 25th January; Discussion in *Youth and Technology* (1961), Nos. 1, 2 and 3; V. P. Tugarin (1961), Thought transference—again, *Knowledge is Power*, No. 7, p. 22.

(11) P. I. Guliacv (1960), *The electrical activity of the human cerebral cortex*, Leningrad: University Press.

(12) J. Bergier (1959), La transmission de pensée—arme de guerre, *Constellation*, No. 140, December: G. Messadié (1960), Du Nautilus, *Science et Vie*, No. 509, February.

(13) One must exercise a certain caution with respect to the contents of these articles as authoritative sources in Washington disclaim all knowledge of such experiments. Meanwhile it is also known that the Laboratory of Parapsychology at Duke University, North Carolina, U.S.A., in 1952 received a financial grant from the U.S. Office of Naval Research for experiments in extrasensory perception: J. B. Rhine and J. G. Pratt (1957), *Parapsychology—Frontier Science of the Mind*, Springfield, p. 203.

(14) L. L. Vasiliev (1960), On the electromagnetic radiation of the brain, paper read at the 15th Conference of Science and Technology, *Age of Wireless*, Journal of the All Union Radio Technological Society.

Notes to Chapter 1

(1) E. Gurney, F. Myers and F. Podmore (1886) *Phantasms of the Living*, London. There is a Russian translation, published in 1893 in St. Petersburg.

(2) See R. Amadou (1954), *La Parapsychologie*, Paris, pp. 81–88.

(3) Ch. Richet (1884), La suggestion mentale et le calcul des probabilités, *Revue Philosophique*, No. 18, p. 609.

(4) Particulars concerning these works are given in Note (1) to Chapter 5.

(5) E. Osty (1925), La Télépathie Expérimentale, *Revue Métapsychique* No. 1, p. 5.

(6) G. Murphy (1925), cited by E. Osty, *Revue Métapsychique* No. 1, p. 13.

(7) E. Bozzano (1933), Considérations et hypothèses au sujet des phénomènes télépathiques, *Revue Métapsychique* No. 3, p. 145.

(8) Terms introduced by Prof. Charles Richet and J. Ochorovicz.

(9) R. Warcollier (1924), La télépathie active et passive, *Revue Métapsychique*, No. 5.

(10) E. Bozzano (1933), Considérations et hypothèses au sujet des phénomènes télépathiques, *Revue Métapsychique*, No. 3, p. 145.

(11) Term introduced by Myers instead of the obsolete and misleading word "clairvoyance," F. Myers (1909), *Human Personality*, New Impression, New York.

(12) A similar hypothesis was advanced even earlier by F. Myers, who considered that we are here dealing with vibrations of the "meta-æther."

(13) See note (7) to this Chapter.

(14) Report in the section on electricity of the Franklin Institute (U.S.A.), 1892, E. Houston (1899), La radiation cérébrale, in the book by A. de Rochas, *L'extériorisation de la sensibilité*, Paris, pp. 215–225.

Notes to Chapter 2

(1) For details see the pamphlet recently published by P. I. Guliaev (1960), *Electrical Activity of the Human Cerebral Cortex,* Leningrad: University Press.

(2) P. P. Lazarev (1960), *Current Problems in Biophysics,* Moscow, pp. 28–29. By the same author (1922): *The physico-chemical basis of higher nervous activity,* Moscow; by the same author (1923): *Russian Journal of Physiology,* Vol. 5, p. 312.

(3) V. M. Bekhterev (1920), Contribution to *Problems in the Study and Training of Personality,* see Note (1) to Preface; by the same author (1921) in *General Reflexology,* Collos Edition, pp. 122–126. 3rd Edition G. E. Z. 1925, p. 38.

(4) F. Cazzamalli (1925), Phénomènes télépsychiques et radiations cérébrales, *Revue Métapsychique,* No. 4, p. 3.

(5) Director of the French General Company of Wireless Telegraphy.

(6) A. A. Petrovsky (1926), Telephysical phenomena and mental radiation, *Wireless Telephony and Telegraphy,* No. 34, p. 61.

(7) N. A. Skritsky and V. V. Lermontov (1926), *First report on the subject of the effect of a human body on receivers and transmitters over the kilocycle range,* Berlin.

(8) A. A. Petrovsky (1926), Telephysical manifestations and mental radiation, *Wireless Telephony and Telegraphy,* No. 34, p. 61.

(9) F. Cazzamalli (1929), *Esperience, argomenti e problemi di biofisica cerebrale,* Geneva.

(10) F. Cazzamalli (1933), *Giornale di Psychiatria e di Neurologia,* Estratto Fasc. 1, I. Trimestre.

(11) The physical part of these experiments was carried out by Gnezoutta, an engineer and collaborator of Marconi's, and by the technician Riza.

(12) According to W. S. Schumann (1928), Ueber elektrische Felder physiologischen Ursprung's, *Zs. f. technische Physik,* Leipzig, No. 3, p. 315.

(13) F. Sauerbruch and W. Schumann (1928), Ueber elektrische Felder in der Umgebung lebender Wesen, *Zs. f. technische Physik,* p. 96.

(14) B. B. Kajinsky (1923), Thought transmission, *News of the Association for Natural Science,* Appendix 2M; B. B. Kajinsky (1962), *Biological Radio Communication,* Kiev, Ed. A.N. U.S.S.R.

A. V. Leontovich (1928), Comptes rendus de l'Académie Française, Paris; A. V. Leontovich (1935), The neuron as apparatus of alternating current, paper at the 15th International Congress of Physiology; G. Lakhovsky, *Les oscillations cellulaires,* Paris.

(15) The same objection was raised by Norbert Wiener (see *Youth and Technology,* 1961, No. 3, p. 30). However, the Polish physicist and parapsychologist Stephan Manczarski reached a different conclusion: according to him a human brain can produce a "wide range of radiations"—from waves of a centimetre to waves of tens of hundreds of kilometres in length: S. Manczarski

(1961), Zastosowanie cybernetyki radiofysiki w parapsychologii, *Przeglad telekommunikacyung,* No. 11, p. 325.

(16) The physiologist Lorente de Nó showed that the anatomical arrangement of neurons in the cerebral cortex makes possible a continuous series of impulses between neurons, which he called a closed self-stimulating chain. Circulating continuously along a specific closed neuron chain the impulses constitute a specific activity, or space-time pattern, which may serve as a partial base for self-similar specific psychic manifestations, for instance thoughts, feelings, acts of will, etc. It is, however, quite possible to argue that such specific patterns of circulating impulses in their turn generate electromagnetic waves that are equally specific in composition, and that these waves may, at a distance, elicit in another brain corresponding spatio-temporal patterns circulating along neuronal chains and initiating similar parallel mental experiences. It should be added that it is possible, by means of contemporary radio techniques, to simulate the "specific patterns of nervous activity" as understood by electrophysiologists by means of specific and complex combinations of electrical impulses and fields: Lorente de Nó (1933), Studies on the structure of the cerebral cortex, I, *J. Psychol. Neurol.,* Vol. 45, p. 381; see also J. C. Eccles (1953), *The neurophysiological basis of mind,* pp. 217–219.

(17) As opposed to *inter*cerebral induction, i.e. transmission from one brain to another (which Academician Leontovich does not study).

(18) V. Arkadiev (1924), Concerning the electromagnetic hypothesis of mental suggestion, *Journal of Applied Physics,* Vol. I, p. 215.

(19) Arkadiev, *loc. cit.* (18), pp. 219–220. It should be noted, however, that it has been shown by recent research that indeed powerful shortwave emissions are not without effect on the psychoneural aspects of personality, and may cause a number of pathological symptoms such as headache, tightness around the temples, insomnia and weakness accompanied by arm and leg tremor. New data on this subject are to be found in the collection of papers edited by K. Triumfov (1957), *The biological effects of ultra-high frequency magnetic fields,* Leningrad.

(20) V. U. Danilevsky (1900), *The physiological effects of electricity at a distance,* Vol. 1, 1900, Vol. 2, 1901, Kharkov; F. P. Petrov (1935), The effect of an electromagnetic field on isolated organs, from: The physico-chemical basis of nervous activity, *Proceedings of the Institute for Brain Research.*

(21) F. P. Petrov (1952), The effect of a low-frequency electromagnetic field on higher nervous activity, *Proceedings of the I. P. Pavlov Institute of Physiology,* A.N. U.S.S.R., Vol. I, p. 369.

(22) U. A. Holodov (1958), Contribution to the study of the physiological analysis of the effects of a magnetic field on animals, Doctoral thesis, Moscow.

(23) M. U. Ulianov (1959), Establishing a temporary experimental connection in an animal's brain by means of electrical stimulation, *Scientific Notices of the Gorky State Medical Institute,* Part 9.

(24) W. Penfield (1956), Mental phenomena effected by means of cortical stimulation, Journal V.N.D. Vol. 6, 4; W. Penfield and H. Jasper (1954), *Epilepsy and the functional anatomy of the human cerebral cortex,* London.

(25) The first report of these experiments appeared in an article by A. L. Chijov (1925), Transmission of thought at a distance (brain-mechanism radio), in the journal *Echo,* No. 20.

(26) The paper was called "On the electromagnetic radiation of man," but it was never published; it is referred to here on the strength of a type-written copy which was preserved.

(27) A brief outline of this paper is sketched in an article by B. V. Kraiuhin (1948). Can electromagnetic induction exist in the tissues of the living organism? *Collection of papers in memory of A. V. Leontovich,* Kiev, p. 88. For further details see S. I. Turligin (1942), Study of microwaves ($\lambda \approx 2mm$) by means of the human organism, *Bulletin of Experimental Biology and Medicine,* Vol. 14, 4, No. 10, p. 63.

(28) For some new considerations concerning the question of the electromagnetic fields of the brain, see an article by R. M. Granovsky (1961), of that title, in the *Proceedings of the Leningrad Society for Experimental Natural Sciences,* Vol. LXXII, I, p. 111.

Notes to Chapter 3

(1) According to information received in 1960, from the Institut Métapsychique International, Paris, the Cazzamalli experiments have remained unconfirmed.

(2) As has already been pointed out in the preceding Chapter, S. I. Turligin obtained the same result.

(3) See Note (25) to Chapter 2, and Note (22) to Chapter 4.

(4) This was written some 30 years ago. Now it is known that long (kilometre) electromagnetic waves penetrate in a weakened state an iron wall 1 mm in thickness. About 80% of the energy of a low frequency electromagnetic field (50 c/sec.) is reflected from the walls of the chamber, the remaining 20% of the field's energy which passes through the 1 mm thickness is weakened approximately 3 times as regards intensity, and 33 times as regards power (data obtained by I. P. Ionov).

Notes to Chapter 4

(1) P. Joire (1897), De la suggestion mentale, *Annales des sciences psychiques,* No. 4, p. 193; No. 5, p. 263; Charles Richet (1886), *Des mouvements inconscients,* Paris; P. Janet (1889) *L'automatisme psychologique,* Paris. There is a Russian translation of this latter book, published in 1913.

(2) Similar experiments with positive results on mental suggestion of definite movements were conducted by Dr. N. Krainsky (1900), *Damage: Hysterical Women and Demoniacs,* Novgorod.

(3) A description of this phenomenon is given in Note (4) below.

(4) If one presses on a portion of the skin covering certain nerves, the underlying muscles will contract and remain in a state of stable contraction. Thus, for example, pressure on the radial nerve will cause stretching of all the fingers of the hand, stimulation of the median nerve will cause all the fingers to contract; if the ulnar nerve is stimulated, the second and third fingers are extended, the first and fourth bending towards the palm of the hand. Even if the hypnotist does not know what to expect as a result of such stimulation, these reflex actions occur regularly. However, pressure on the same nerves, in the case of the same experimental subject is without effect if the subject is awake. This "Charcot effect" is not of peripheral neuromuscular, but of central origin. We were able to establish quite easily in the hypnotised subject Kouzmina a "neuro-muscular hyper-stimulation reflex" in response to a given signal, e.g. sound. To establish such a reflex it is necessary to accompany the pressure on the nerve (an unconditioned stimulus) by a sound signal (the conditioned stimulus); even after the application of a small number of this combination of stimuli the signal alone elicits the reflex as strongly as if mechanical stimulation had been applied. For further details of neuro-muscular hyper-stimulation see Charles Richet (1886), *Clinical Essay on Grand Hysteria*, Kharkov, 299–305.

(5) The experiments of 6/10/1926 were made on account of the arrival in Leningrad of a well known physiologist, A. A. Kuliabko, who wanted to assess the reality of mental suggestion.

(6) For details see A. I. Bronstein (1930), Registering the motor acts of human subjects, in Problems in the Study and Training of Personality, *Proceedings of the Institute for Brain Research,* I and II, Leningrad, p. 98.

(7) For this type of experiment an electromagnetic recording device set at the beginning of suggestion is not recommended because the recorder, when switched on or off, makes a slight noise which might be heard by the subject and act as a signal.

(8) L. L. Vasiliev and G. U. Belitzky (1944), Influencing ideomotor reactions, *Bulletin of Experimental Biology and Medicine,* Vol. 17, I–II, p. 26.

(9) E. Jacobsen (1930), Electrical measurement of neuromuscular states during mental activities, *Am. J. Physiol.,* Vol. 91, p. 567.

(10) M. S. Bikhov, Experiments in the electrophysiological analysis of motor recordings in the light of Pavlov's teaching, Dissertation for doctoral thesis, Leningrad University.

(11) The actual kymograms of these experiments have not been preserved by the author; on account of their age only the descriptions have been kept, and these are here reported.

(12) H. Brugmans, Quelques expériences télépathiques faites à l'Institut Psychologique de l'Université de Groningen, *Compte rendu officiel du I.* Congrès International des recherches psychiques, Copenhagen, 1922, p. 397.

(13) In approximately half of the 187 experiments the experimenters were in the same room as Van Dam, but in these cases the results were poorer than when they were in the upstairs room.

(14) The probability is calculated by the formula $(1/48) \times 187 = 3.89$ (4 in round numbers).

(15) V. M. Bekhterev (1920), Experiments on the effects of "mental" influence on the behaviour of dogs, *Problems in the Study and Training of Personality,* Petrograd, 2nd Edition, p. 230.

(16) A. G. Ivanov-Smolensky (1920), Experiments in mental suggestion on animals, *Problems in the Study and Training of Personality,* Petrograd, 266.

(17) A. G. Ivanov-Smolensky himself gave further careful consideration to his experiments; from his point of view the two experiments referred to above were only "partially successful," the problem itself remaining unsolved but warranting further investigation.

(18) A. Ferrière (1948), Expériences de télépathie chez des chiens et chez des humains (Extrait du traité de Bekhterev, *Réflexologie Collective,* Petrograd), *Revue de Métapsychique,* No. 3, p. 165.

(19) J. B. Rhine (1949), in his *Journal of Parapsychology,* Vol. 13, p. 166 seq. reprinted the above-mentioned article by V. M. Bekhterev, "Experiments on the effects of 'mental' influence on the behaviour of animals."

(20) According to the article ". . . . pokes her nose in it," in V. M. Bekhterev (1920), *loc. cit.,* p. 230.

(21) Sub-sensory or sub-threshold reactions: those responses to the stimulation of receptor organs which reach the cerebral cortex but which are not consciously noted owing to insufficient intensity of the stimuli. It is, however, possible to establish conditioned reflexes in response to such unnoticed responses, see G. V. Gershouni *et al.* (1948), Properties of temporal associations with unnoticed auditory stimulation, *Bulletin of Experimental Biology and Medicine,* 26, pt. 3.

(22) B. B. Kajinsky (1928), Das Tier im Banne der Gedanken, *Wissen und Fortschritt,* 12, pp. 262–267; B. B. Kajinsky (1962), Biological radio-communication, Kiev A.N. U.S.S.R.

(23) This work was reported in 1937 to the meeting of the Moscow Society of Psychologists and Neuropathologists, but the data were not published. I am here using a typewritten copy of the text for the report.

(24) Mentioned in an unpublished article by S. I. Turligin, "The emission of electromagnetic waves by the human organism," (p. 71 of the typescript). This article was published in abridged form in the *Bulletin of Experimental Biology and Medicine* (1942), Vol. 14, 4, 10, pp. 63–72.

(25) As has already been mentioned in this Chapter, we employed this mental suggestion, i.e. to fall backwards, in the middle of the 'thirties, but we used the Bronstein pneumatic platform for the purpose of registering the subject's responses.

(26) S. Alrutz (1922), *Compte rendu du I. Congrès international des recherches psychiques,* Copenhagen, p. 278. For details see S. Alrutz (1924), *Neue Strahlen des menschlichen Organismus,* Stuttgart.

(27) S. Alrutz (1922), *loc. cit.*

(28) R. Reutler (1928), L'action par distance des organismes vivants: sur les organes vivants isolés, *Revue Métapsychique,* 3, p. 197.

(29) To prevent the drying out of the preparation.

(30) Died in the siege of Leningrad before completing his work. I am basing the data given here on a typewritten report by the author which I have kept.

(31) K. Wachholder und Altenburger (1925). Beiträge zur Physiologie der willkürlichen Bewegungen, *Pflüger's Archiv,* 209, 2–3.

(32) F. Sauerbruch und W. Schumann. Ueber elektrisehe Felder physiologischen Ursprungs, *Zs. f. techn. Physik,* 3, pp. 96 and 315, Leipzig.

Notes to Chapter 5

(1) W. Wasilewsky (1925), *Telepathie und Hellsehen,* Stuttgart; R. Tischner (1925), *Telepathy and Clairvoyance,* London; R. Warcollier (1926), *La télépathie expérimentale,* Conf. de l'Inst. Métapsychique International, Paris; U. Sinclair (1930), *Mental Radio,* Pasadena Station, U.S.A. W. Prince (1932), *The Sinclair experiments demonstrating telepathy,* Boston, E. Osty (1932), Télépathie spontanée et transmission de la pensée expérimentale, *Revue Métapsychique,* Nos. 4, 5, 6; S. G. Soal (1932), Experiments in supernormal perception at a distance, *Proc. Society for Psychical Research,* Vol. 40; J. B. Rhine (1934), *Extrasensory Perception,* Boston.

(2) Among older works of Russian authors the experiments on mental suggestion of drawings conducted by J. Jouck, a lecturer, deserve special mention: J. Jouck (1902), The reciprocal connection between organisms, God's Universe, No. 6; also the monograph by Dr. N. G. Kotik (1912), *Direct thought transmission, an experimental survey,* Moscow.

(3) The Second All-Russian Psychoneurological Congress took place in January 1924 in Petrograd. The Proceedings of the Congress were not published. Some details of the work reported at the meeting are to be found in Appendix A to this book.

(4) The report was compiled and read at the meeting by V. A. Podierny at the instance of the Commission for the Study of Mental Suggestion. A typewritten copy has been preserved among the papers of L. L. Vasiliev and will be referred to later.

(5) See: L. L. Vasiliev (1959), *Mysterious phenomena of the human psyche,* Chapter V, Moscow, State Political Publishing House.

(6) The members of the Commission for the Study of Mental Suggestion acted as senders; a number of different subjects acted as percipients and it is not now possible to ascertain their names.

(7) On each Zener card there is a distinct picture of one of five different figures: circle, rectangle, cross, star, three wavy lines.

(8) R. Warcollier (1929), La télépathie, ses rapports avec l'inconscient et l'inconscient, *Revue Métapsychique,* No. 4, p. 270.

(9) R. Desoille (1932), De quelque conditions auxquelles il faut satisfaire pour réussir des expériences de télépathie provoquée, *Revue Métapsychique,* No. 6.

(10) Some authors make a distinction between unconscious and subconscious mental experiences, the unconscious ones being those that cannot under any circumstances be recalled in memory, e.g. experiences of very early childhood; the *sub*conscious [or *præ*conscious] ones being those which, under certain circumstances, can be brought into consciousness, though often in symbolic form. Other authors consider such a distinction to be superfluous. I consider it to be more accurate to make such a sub-division.

(11) F. Mikhailov and G. Zaregoyrodzev (1961), *Beyond the threshold of consciousness,* a critical analysis of Freudianism, Moscow, p. 45, author's italics.

(12) E. S. Airapetianz and K. M. Bikov (1942), The study of interoception and the psychology of the subconscious, *Achievements of Modern Biology,* Vol. XV, 3, p. 275.

(13) Similar examples are given, under the name "telepathy of latent memory" in the report of Dr. Tanagra (1930), La télépathie de la mémoire latente, *Transactions of the Fourth International Congress for Psychical Research,* London, 1930, p. 837.

(14) A description of this method by Academician Mitkevich himself is given in Appendix C of this book.

(15) Charles Richet (1884), La suggestion mentale et le calcul des probabilités. *Revue Philosophique,* No. 18, p. 609.

(16) $1 : 2^{10} = 1 : 1024$, i.e. one success in 1024 experiments.

(17) R. A. Fisher (1958), *Statistical methods for experimenters,* Moscow, p. 142.

Notes to Chapter 6

(1) P. Janet et M. Gibert (1886), Sur quelques phénomènes de somnambulisme, *Revue Philosophique,* I et II, 1886; P. Janet (1889), *L'automatisme psychologique,* Paris, Alcan, p. 103; F. Myers (1909), *Human Personality,* London, pp. 382–383; J. Ochorovicz (1887), *De la suggestion mentale,* Paris. Myers and Ochorovicz also took part in these experiments.

(2) See Note (26) to Chapter 4.

(3) B. N. Birman (1925), *Experimental Sleep,* Leningrad.

(4) Subsequently, see Chapter 7, we shall call this series of experiments the initial version of the hypnogenic method.

(5) In the experiment carried out with the registering apparatus in the same room as the subject the time of beginning of mental suggestion and

awakening was marked on the kymograph by the experimenter drawing a line on the record. The reason for this was that the switching on of the recorder into the circuit makes a slight noise, which could have been heard by the subject.

(6) H. Brugmans (1922), *Compte Rendu Officiel du I. Congrès International des Recherches Psychiques,* Copenhagen, p. 397.

Notes to Chapter 7

(1) The thickness of the walls of our iron chambers was about 1 mm (see Chapter 2).

(2) This technique is essentially analogous to that subsequently employed by S. I. Turligin who also, in his experiments, did not produce, but merely speeded up by mental suggestion, a reaction which the experimental conditions were designed to evoke in any case—that of the subject falling backwards.

(3) Metre, centimetre and millimetre waves were completely absorbed by both iron and lead chambers. As regards the exclusion of long waves (kilometre waves) the lead chamber, despite the fact that its walls were 3 times thicker, was hardly more effective than the iron chambers.

(4) That this was an autohypnotic condition and not ordinary sleep is shown by the fact that the sleeping subject talked to the sender, Tomashevsky, when he entered the chamber after having received a signal to the effect that the percipient had fallen asleep, and that she replied to his questions.

Notes to Chapter 8

(1) B. N. Birman (1925), *Experimental Sleep,* State Publishing Office, Leningrad.

(2) G. V. Gershuni (1947), Researches on sub-threshold reactions in the activity of sensory organs, *Physiol. 1. U.S.S.R.,* Vol. 33, p. 393.

(3) See the book by Gurney, Myers and Podmore, *Phantasms of the Living,* cited in Chapter 1.

(4) R. Warcollier (1927), La télépathie à très grande distance, *Compte rendu du III. Congrès International des recherches psychiques,* Paris.

(5) Upton Sinclair (1930), *Mental Radio,* Pasadena Station, U.S.A.

(6) K. K. Konstantinides (1930), Telepathische Experimente zwischen Athen, Paris, Warschau und Wien, *Transactions of the IVth International Congress of Psychical Research,* Athens.

(7) E. Bozzano (1933), Considérations et hypothèses au sujet des phénomènes télépathiques, *Revue Métapsychique,* No. 3, p. 145. Prof. Stephen Manczarski also finds that the best results of experiments in mental suggestion are obtained at a distance of up to 4 m (see Note 15, Chapter 2).

(8) W. Barrett (1914), *Puzzling phenomena of the human psyche,* Moscow.

Notes to Chapter 9

(1) E. Osty (1925), La télépathie experimentale, *Revue Métapsychique,* No. 1, p. 5.

(2) Caslian, cited by R. Warcollier (1926), La télépathie expérimentale, *Les Conférences de l'Institut Métapsychique,* Paris, p. 48.

(3) The same was observed by us in the experiments in mental suggestion of acts of movement with subject Kuzmina (see Chapter 4).

(4) "Working" means the sender is mentally suggesting "fall backwards."

(5) S. I. Turligin (1942), Radiations of microwaves ($\lambda \approx 2mm$) by the human organism, *Bulletin of Experimental Biology and Medicine,* Vol. 14, 4, No. 10, pp. 66–67.

(6) We will refer to this again in the next Chapter.

(7) According to Norbert Wiener an aerial of the size of the human body can emit electromagnetic waves only at a frequency of a million oscillations per second. Therefore the assumption that the low frequency of the alpha-rhythm can be emitted with sufficient intensity by a body is incorrect. One must search for high-frequency radiations of the brain. (This question has not been studied sufficiently, see editorial article in *Youth and Technology,* 1961, p. 30.)

(8) These conclusions were formulated by us in the report of the work written in 1934.

(9) K. I. Platonov, *The word as physiological and medical factor,* Moscow, 1930, 3rd edition, 1962.

Notes to Chapter 10

(1) *Report on Five Years of Activities (1954–1958) of the Parapsychology Foundation, Inc.,* New York, U.S.A., 1959.

(2) J. B. Rhine and J. G. Pratt (1957), *Parapsychology,* Springfield, U.S.A.; R. Amadou (1954), *La Parapsychologie,* Edit. Danoël; B. M. Humphrey (1948), *Handbook of Tests in Parapsychology,* Durham, U.S.A.

(3) Prof. Vasiliev should here have referred to the Society for Psychical Research, London.

(4) B. F. Skinner (1948), Card-guessing experiments, *American Scientist,* Vol. 36, No. 3, p. 458; E. G. Boring (1955), The present status of parapsychology, *American Scientist,* Vol. 43, No. 1, p. 108; C. E. M. Hansel (1959), Experiments on telepathy, *New Scientist,* Vol. 5, February; C. E. M. Hansel (1959), Experimental evidence for extra-sensory perception, *Nature,* November, Vol. 184, No. 4697.

(5) These experiments are analysed in detail and some grave errors are pointed out in the book by S. G. Soal and F. Bateman (1954), *Modern Experiments on Telepathy,* New Haven, Yale University Press, p. 15.

(6) G. Price (1955), Science and the Supernatural, *Science,* No. 3165, Vol. 122, p. 389.

(7) G. Murphy (1957), Thought transference is a fact, *This Week,* February.

(8) G. Price, *loc. cit. Science.* Here as well as later on, when quoting the contents of the article by Price, I employ his terminology, using inverted commas.

(9) Price's article did not remain unanswered by parapsychologists, in particular by Rhine and Soal (see controversies in the *Journal of Parapsychology,* 1955, Vol. 19, No. 4, pp. 236–271).

(10) A similar criticism of the achievements of parapsychology, but less serious, is contained in a book by D. H. Rawcliffe (1952), *The Psychology of the Occult,* London.

(11) This is what used to be called, and is still nowadays often called, clairvoyance in space. One must note that the term clairvoyance was used by Pavlov. For example, he wrote: ". . . in some instances, during normal conscious activity, a person's ability to differentiate becomes more acute. Under special conditions of so-called clairvoyance the ability of man to differentiate becomes infinitely acute" (I. P. Pavlov, *Complete Collected Works,* Moscow and Leningrad, 1952, p. 520). I. P. Pavlov certainly attributed such clairvoyance to the usual sense organs, whereas the great French physiologist Charles Richet and the modern parapsychologists attribute clairvoyance to a still unknown form of sensitivity which they call concealed sensitivity—cryptæsthesia or telæsthesia.

(12) J. B. Rhine and J. G. Pratt (1957), *Parapsychology,* Springfield, pp. 54–56.

(13) Instead of using the customary Zener cards with abstract symbols (circle, cross, star, square, wavy lines) Soal often used cards with pictures of animals (lion, elephant, zebra, giraffe, pelican).

(14) S. G. Soal and F. Bateman (1954), *Modern Experiments in Telepathy,* Yale University Press.

(15) In the old terminology: "clairvoyance in time" as distinct from telæsthesia—"clairvoyance in space."

(16) R. Kherumian (1949), Essai d'interprétation des expériences de Soal et Goldney, *Revue Métapsychique,* No. 8, p. 222.

(17) The principal work of Frederic Myers, *Human Personality and its Survival after Bodily Death,* London, is characteristic in this respect.

(18) However, there are still differences of opinion concerning this question. For example, Rhine asserts that a good percipient can perceive a telepatheme from any person. Soal, on the contrary, attributes great importance to the selection of telepathic pairs.

(19) See for instance G. Schmeidler (1960), *E.S.P. in relation to Rorschach Test evaluation,* Parapsychology Foundation Inc., New York.

(20) The inexplicable nature of parapsychical phenomena is discussed in detail in a voluminous article by R. Kherumian (1948), L'introduction à

l'étude de la connaissance parapsychique. *Revue Métapsychique,* Nos. 1, 2, 3. In this article the writer cites with approval the work of E. S. Airapetianz and K. M. Bikov (1946), L'expérimentation physiologique et la psychologie de l'inconscient, published in the journal *La Revue Internationale,* No. 8, p. 116. In that article an effort is made to approach the problem of the subconscious from the point of view of the conditioned reflex school. See also the article by the same authors, "The study of interoception and the psychology of the subconscious," *Achievements of Modern Biology,* Vol. XV, No. 3, 1942, p. 273, already cited in Chapter 5.

(21) M. Ryzl (1962), Training of E.S.P. by means of hypnosis, *Journal of the Society for Psychical Research,* March, No. 711.

(22) R. Warcollier (1928), L'accord télépathique, *Revue Métapsychique,* p. 24.

(23) R. Warcollier et R. Kherumian (1954), Remarques sur la déformation des dessins télépathiques, *Revue Métapsychique,* 27, p. 149.

(24) *Report on Five Years of Activities (1954-1958) of the Parapsychology Foundation, Inc.,* of New York, 1959, p. 42.

(25) N. G. Louwerens (1960), E.S.P. experiments with nursery school children in the Netherlands, *Journal of Parapsychology,* Vol. 24, 2, p. 75.

(26) M. de Thy (1959), Télépathie et déficience mentale, *Revue Métapsychique,* Vol. 2, No. 10, p. 4.

(27) For example, in the American *Journal of Parapsychology* (1949), Vol. 13, p. 166, an article appeared, called "V. M. Bechterev: Direct influence of a person upon the behavior of animals."

(28) L. J. Franke et L. J. Koopman (1939), Le fonctionnement du cerveau humain pendant des phénomènes métapsychiques, extrait de la *Revue Métapsychique,* Janvier/Février.

(29) For a popular description of the methods of electro-encephalography and a description of the bioelectric potentials of the brain, see P. I. Gouliaev (1960), *Electrical processes of the human cerebral cortex,* Leningrad University.

(30) M. N. Livanov and V. M. Ananiev (1959), Electroencephaloscopy, *Medical Journal,* Moscow.

(31) See Chapters 4 and 5 of this book.

(32) J. L. Woodruffe and L. A. Dale (1952), E.S.P. function and the psychogalvanic response, *Journal of the American S.P.R.,* Vol. 46, p. 62.

(33) S. Figar (1959), The application of plethysmography to the objective study of so-called extra-sensory perception, *Journal of the Society for Psychical Research,* Vol. 40, No. 702, p. 162.

(34) D. J. West (1959), Comments on Dr. Figar's paper, *Journal of the Society for Psychical Research,* Vol. 40, No. 702, p. 172.

(35) R. Warcollier (1954), L'antagonisme entre les images mentales et le problème des écarts négatifs en ESP, *Revue Métapsychique,* Nos. 29–30, p. 54

(36) R. Kherumian (1959), Réflections sur l'état actuel et les perspectives de la parapsychologie, *Revue Métapsychique,* Vol. II, No. 9, p. 4.

(37) From a review of the work of P. A. Castruccio, E.S.P. in the Industrial Research Laboratory, *Parapsychology Bulletin* (1960), August, No. 54.

(38) R. Dufour (1951), L'espace, joue-t'il un rôle dans les phénomènes de télépathie? *Revue Métapsychique,* No. 13, p. 2 and No. 14, p. 49.

(39) J. B. Rhine and J. G. Pratt (1957), *Parapsychology,* Springfield, pp. 66–9.

(40) W. Carrington (1948), *La télépathie,* Paris.

(41) B. Heffman (1940), Extrasensory perception and the inverse square law, *Journal of Parapsychology,* Vol. 4, No. 1.

(42) I. A. Poletayev (1958), *Signal,* Moscow, p. 24.

(43) D. A. Biriukov (1959), *The myth of the soul,* Moscow, Edn. Soviet Russia, pp. 129–152.

(44) H. Saradiev considers that "the perception of the functionings of one brain by another brain must be effected through some receptor sense organs, but not directly from brain to brain. It is possible that such receptors may be embedded in the skin." In support of this he cites some interesting facts. Some fish with under-developed electrical organs orientate themselves in muddy water by means of electrical receptors which perceive the changes of the electromagnetic field in the surrounding water generated by the discharge of another fish. The electro-receptors are in the side of the fish and capable of reacting to changes in the potential of the field of only 3.10^{-9} volts per 1 mm^2. In this instance through 1 cm^2 of the surface of the skin of the body there flows a current of 2.10^{-11}, whereas a current 100,000 times stronger is necessary in order to cause a stimulation of the nervous fibrils (H. Saradiev, 1961, Can a brain perceive transmission of thoughts independently? *Youth and Technology,* No. 1, p. 31).

(45) R. Kherumian (1958), Réflexions sur l'état actuel et les perspectives de la parapsychologie. *Revue Métapsychique,* Décembre, No. 8, p. 4.

(46) I. A. Fabri (1935), Observations and experiments on the sexual life of *Saturnia puri* Schif *(Lepidoptera), Entomological Review,* Vol. XXV, Nos. 3–4, p. 314.

(47) G. P. Frolov (1959), Puzzles of Scent, *Youth and Technology,* No. 12, pp. 27–28.

(48) For example, Prof. S. Manczarski, in his work already cited (Chapter 2, Note 15), asserts such a possibility. In his opinion, the transmission of the telepatheme through the walls of a metal chamber is effected by kilometre electromagnetic waves which are produced by the sender's brain. It is now known that electromagnetic waves of sufficient length also penetrate sea water, for instance if they are 30 km or more. The intensity of the waves is then weakened approximately 17 times to every 10 m of sea water.

(49) V. K. Volkers and W. Candib, Detection and analysis of high frequency signals from muscular tissues with ultra-low-noise amplifiers. Paper presented at *International Convention of Radio Engineers,* quoted in *Newsletter of the Parapsychology Foundation, Inc.,* 1960, Vol. 7, No. 2, p. 1.

(50) R. Kherumian (1949), Essai d'interprétation des expériences de Soal et Goldney, *Revue Métapsychique,* No. 8, p. 223.

(51) H. Berger (1940), *Psyche;* W. C. Roll (1960), *"Psyche* by Hans Berger," book review, *Journal of Parapsychology,* Vol. 24, No. 2.

(52) In this connection the discussion which took place in 1930 between Academician V. F. Mitkevich and J. I. Frenkel is of interest for parapsychologists. Mitkevich supports the point of view of Faraday and Maxwell, according to which physical action at a distance can only occur through a medium in which there are material particles of some sort, even æther; from this point of view action at a distance through completely empty space is impossible. Frenkel, on the other hand, points out that an electromagnetic field can spread even through completely empty space. See Mitkevich (1930), *On "physical" action at a distance,* Edn. A.N. U.S.S.R., Leningrad. Parapsychologists who would introduce the concept of a "meta-æther" apparently support the position of Academician Mitkevich, which is however repudiated by most contemporary physicists.

(53) See Note (10) to this Chapter.

(54) R. Kherumian, Réflexions sur l'état actuel et les perspectives de la parapsychologie, *Revue Métapsychique,* Décembre, No. 8.

Note to Appendix A

(1) Apparently this is A. G. Kozhevnikov, Professor of Zoology at Moscow University, who took an active part in the experiments in mental influence on V. L. Durov's trained dogs [my note—L. L. V.]

Note to Appendix C

(1) V. F. Mitkevich's italics.

Notes to Appendix D

(1) J. Ochorovicz (1887), *De la suggestion mentale,* Paris.

(2) Ochorovicz, *loc. cit.* (1887).

(3) A. Binet (1893), *Revue des Deux Mondes,* 15th March, p. 443.

(4) F. Myers (1909), *Human Personality,* New Impression, p. 382.

(5) See C. Flammarion (1900), *L'Inconnu et les problèmes psychiques,* Paris. Russian translation 1901, St. Petersburg.

(6) For a more detailed summary of the literature on hypnotising at a distance: J. C. Roux (1893), Expériences françaises récontes de suggestion mentale, de clairvoyance et d'hypnotisation à distance, *Annales des Sciences Psychiques,* Juillet/Aôut; F. Podmore (1915), *Apparitions and Thought Transference,* London and New York: C. Richet (1923), *Traité de Métapsychique,* 2nd edn. Paris, pp. 25–26, 126–128.

Notes to Appendix E

(1) Prof. Platonov sent this article at my request for publication in this Appendix [my note—L. L. V.].

(2) I referred to this demonstration of experiments in mental suggestion of sleeping and awakening on the subject Mikhailova in Chapter 6 [my note—L. L. V.].

Illustrations

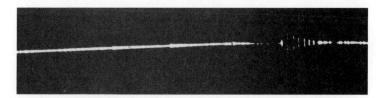

Fig. 1: Oscillogram of brain radio waves according to Cazzamalli.

Fig. 2: Faraday Chamber. There is a bed for the percipient inside the chamber; a radio receiver is on the table.

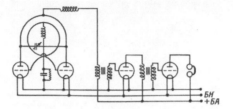

Fig. 3: The scheme of the radio receiver (by Hartley) used in Cazzamalli's and the author's experiments. БH: filament battery; БA: anode lead from high tension battery.

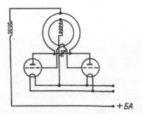

Fig. 4: Short wave generator used in the author's experments. БA: anode battery.

Fig. 5: Our "complete" screening chamber for the sender. Its upper part is raised by a pulley. Inside is a radio transmitter (see Fig. 3). On the right stands the manometer for testing the hermeticality of the chamber when closed, i.e. when the upper half is lowered.

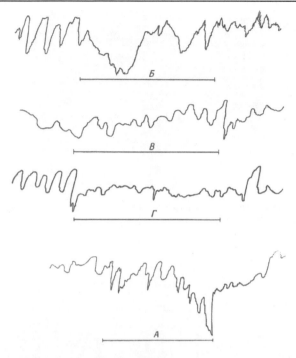

Fig. 6: Characteristics of the fluctuations of ideomotor activity (body sway) of different subjects standing on the pneumatic platform designed by A. I. Bronstein. A line under the kymogram indicates the period of verbal suggestion "sway backwards, etc." Explanation in the text.

Fig. 7: Outer appearance of the amplifier of low frequency radio waves generated by human contracting muscle. Apparatus by R. I. Skariatin, after Sauerbruch and Schumann.

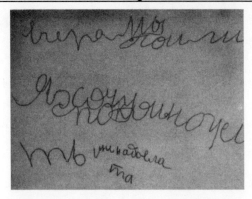

Fig. 8: Automatic script of percipient C. ["Yesterday we went. . . . I want to live peacefully. . . . I am fed up with . . . ta"].

Fig. 9: Automatic script of percipient C., to whom *the number 8* was mentally suggested.

Fig. 10: Automatic script (on the right) of C. when the astronomical sign of the planet Earth and the letter "y" (shown on left) were mentally suggested to her.

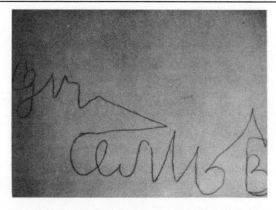

Fig. 11: Automatic script of percipient C. "one and seven = 8" when *the figure 8* was suggested to her for a second time ["one ... seven ... 8"].

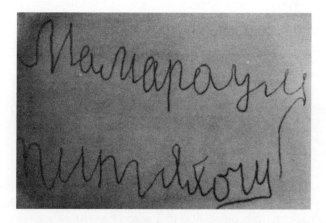

Fig. 12: Automatic writing of percipient C. in response to mental suggestion of "Elma," the name of a deceased young girl ["Mamara ... um (die) ... sleep I want"].

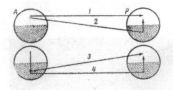

Fig. 13: Schema of transmission of mental suggestion from sender of agent A to percipient P, 1–4 four types of Desoille. Light half of circle: consciousness; darkened half of circle: subconscious.

Fig. 14: Academician Mitkevich's rotating discs, used in experiments of mental suggestion of "white or black." On the left a disc is shown the rotation of which prevents the experimental assistant from concentration either on white or black (see Appendix C).

Fig. 15: Kymogram of the experiment in which the subject compresses a rubber balloon; repeated mental putting to sleep. 3: repeated mental suggestion to sleep; Π: mental suggestion of awakening the percipient; Γ: onset of hypnotic state.

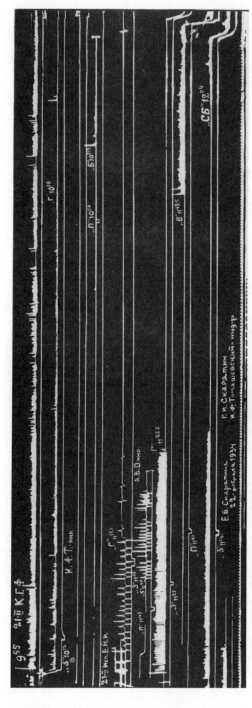

Fig.16: Kymogram of three experiments in mental induction of sleep and wakefulness, carried out on February 21 and 22, 1924 with three subjects. From top to bottom: К.Г.ф. = Federova; Е.М.И. = Ivanova; and Е. С. = E.S. Senders: Tomashevsky = И.ф.Т. and Dubrovsky = A.B.D.; З = go to sleep; П = wake up; Б = state of being awake; C = subject wakes up spontaneously. The figures: time in hours and minutes when one or another condition sets in.

Fig. 17: Experiment with subject A. A. Simultaneous registration of compression of the balloon, the galvanogram obtained by Veragulat's method, and time of onset of mental suggestion to go to sleep. Key same as Fig. 15.

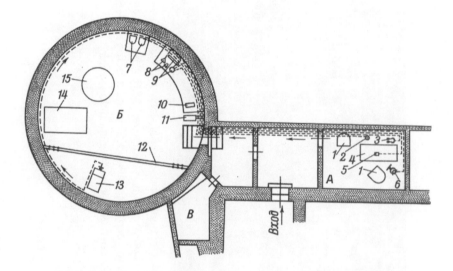

Fig. 18: Plan of the laboratory showing the original positions of equipment (experiments carried out 1933–34). A = "small room" separated from the round room by two passages. In room A there was a couch, 4; two armchairs, 1, for the experimenters; a pneumograph, 5; a microphone, 6; contacts (electrodes) to the mirror galvanometer, 3; and the wiring (signalization) to the kymograph, 2, which were installed in the round room. Б = "round room" (this room is always referred to as room B in the text, the Russian Б being the equivalent of B); Faraday chamber, 14; and "complete" screening chamber, 13. Here, at first, were the registration devices; the kymograph with the Marya pneumatic device, 9; the galvanometer and the shunts for them, 7; the loudspeaker and the telephone, 8, connected with the distant room, A. At a later date, beginning in 1935, all recording apparatus was transferred to room B (this is referred to in the text as room V, the Russian B being the equivalent of V). Amplifier, 10; batteries, 11. Lecker system, 12; working table, 15. The signaling leads are marked in by dashes and arrows.

Fig. 19: Mental sending to sleep and awakening of subject А.И.К. who came to the laboratory for the first time and had never previously been a subject in such experiments. At the top: registration of the compression of the balloon; the lower line: registration of time in minutes; in the center: beginning of mental suggestion to go to sleep. П = wake up.

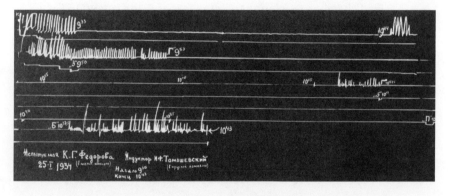

Fig. 20: Registration of spontaneous onset of sleep (autohypnosis), and awakening of subject К.Г.ф. (K.G.F.), Fedorova. Key as Fig. 15.

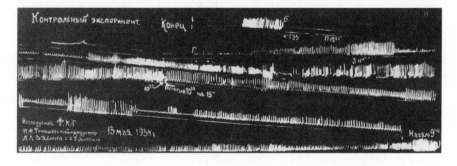

Fig. 21: Prevention of incidence of autohypnosis by carrying on conversation between subject Federova and observer. Control experiment. Kymograph should be read from right to left and bottom to top.

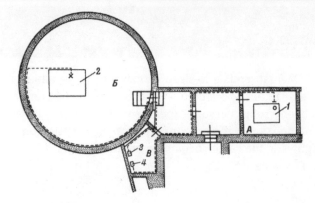

Fig. 22: Plan of laboratory with new set-up of equipment. Room A: 1 = lead chamber. Б (referred to in the text as B, see legend to Fig. 18): room of percipient; 2 = Faraday chamber. В (referred to in the text as V, see legend to Fig. 18): 3 = kymographic installation for recording of sleep or wakefulness of percipient; 4 = button for switching on electromagnetic signal in sender's lead chamber.

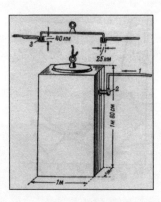

Fig. 23: Outer appearance of the lead chamber for screening the sender. On top there is a schematic representation of the device for hermetically sealing the lid. 1 = leads from recording room; 2 = electromagnetic buzzer coil; 3 = mercury.

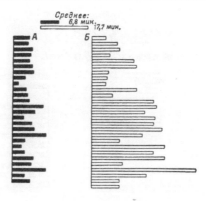

Fig. 24: Diagrammatic representation of results. A = time of falling asleep, speeding-up time of autohypnosis of percipient by means of mental suggestion to go to sleep. Б = time taken by the same percipient to fall asleep without mental suggestion to go to sleep. Average: 6.8 mins. and 17.7 mins., respectively.

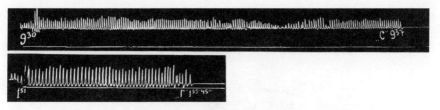

Fig. 25: Kymogram of a separate experiment on sleeping carried out with the same subject: *without* mental suggestion (top), and *with* mental suggestion (bottom). C = sleep, Г = hypnosis.

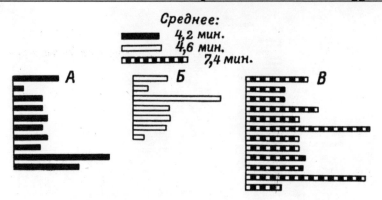

Fig. 26: Diagrammatic representation of results in experiments to accelerate the onset of autohypnosis by means of mental suggestion to go to sleep. A = time taken to go to sleep with mental suggestion and screening. Б = same without screening. B = time taken to go to sleep without mental suggestion. Average: 4.2 mins., 4.6 mins., and 7.4 mins.

Fig. 27: Appearance of the radio set-up generating short wave and radio signals when rubber balloon is compressed by percipient.

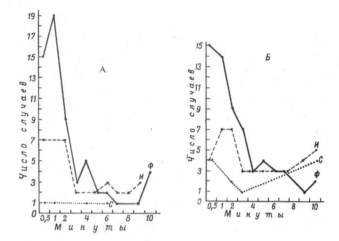

Fig. 28: Curves comparing distribution of lengths of time taken to put to sleep (A), to wake up (Б) by means of mental suggestion of the three subjects ф., И., and С. (F, I., and C.). On the abscissa: time in minutes; on the ordinate: frequency of incidence.

Hampton Roads Publishing Company

. . . for the evolving human spirit

Hampton Roads Publishing Company
publishes books on a variety of subjects,
including metaphysics, health, integrative medicine,
visionary fiction, and other related topics.

For a copy of our latest catalog, call toll-free
(800) 766-8009, or send your name and address to:

Hampton Roads Publishing Company, Inc.
1125 Stoney Ridge Road
Charlottesville, VA 22902

e-mail: hrpc@hrpub.com
www.hrpub.com